SECOND EDITION

Laboratory Handbook for General Chemistry

Norman E. Griswold

Nebraska Wesleyan University

H. A. Neidig

Lebanon Valley College

James N. Spencer

Franklin and Marshall College

Conrad L. Stanitski

University of Central Arkansas

BROOKS/COLE

™

THOMSON LEARNING

Austrailia · Canada · Mexico · Singapore · Spain · United Kingdom · United States

BROOKS/COLE

———✦———™

THOMSON LEARNING

Sponsoring Editor: Bruce Thrasher
Marketing Representative: Tom Ziolkowski
Production Editor: Stephanie Andersen
Production Service: Blue Room Graphics
Manuscript Editor: David Foss
Permissions Editor: Sue Ewing
Interior Design: John Edeen

Cover Design: Christine Garrigan
Cover Photo: Photodisc
Interior Illustration: Brian Betsill
Print Buyer: Kristine Waller
Typesetting: Blue Room Graphics
Printing and Binding: Webcom Limited

Printed in Canada

10   9   8   7   6   5   4   3   2

ISBN: 0-534-97694-8

To the Student

The general chemistry laboratory is the designated place for you to do approved experiments that illustrate and clarify the chemical principles you learn in the classroom portion of the course. The *Laboratory Handbook for General Chemistry* (LHGC) can help you do your laboratory work more effectively, efficiently, and safely.

The LHGC is not a compilation of experimental procedures. Rather, it is a "how-to" guide containing specific information about the basic equipment, techniques, and operations that you will use in your laboratory experiments. The importance of laboratory safety is stressed above all.

This laboratory guide intentionally concentrates on practical matters. More in-depth coverage and theoretical discussions can be found in other sources, including the descriptions in the experiment to be done.

You can use this handbook as:

- a preview of the techniques and manipulations you will use in your upcoming laboratory experiment;

- a review of material covered earlier in the term or in a previous course;

- a handy guide in the laboratory while doing an experiment;

- a reference source for useful data.

You can also use the LHGC effectively with locally written experiments or other laboratory manuals. To help you use this guide, notice that:

- the detailed table of contents makes it easy to find specific topics;

- the appendices contain commonly used reference tables and other useful laboratory information;

- a thorough index helps locate specific items quickly.

The following significant changes have been made in the second edition of the LHGC.

- As often as possible, illustrations and their related text now appear on the same page or facing pages.

- Illustrations have been redrawn to better represent laboratory equipment and operations.

- The order of topics has been revised. Laboratory techniques used early in the first course, such as weighing, measuring liquid volumes, and transferring materials have been moved to the front of the LHGC.

- Questions for review and testing appear at the end of each major section.

- Instructions for using a pH meter have been added.

- A chapter on mathematical operations has been added, including new sections on exponential notation, rounding off numbers, and the equation and slope of a straight line. In addition, the discussion of significant figures has been expanded.

• The graphing chapter has been enlarged to introduce plotting programs to construct graphs. It also includes a graphing exercise involving both hand drawing and use of a spreadsheet program.

We hope you will use the LHGC often and find it a valuable supplement to your laboratory studies.

Contents

I

Safety in the Laboratory

Safety in the chemistry laboratory involves the recognition of all potential hazards and an awareness of the necessary precautions connected with these hazards. Each person working in a laboratory is responsible for the safety of everyone present. Immature behavior has no place in a chemistry laboratory. The potential dangers posed by such behavior are too great.

An accident in a chemistry laboratory can seriously injure or even kill. However, hazards can be anticipated, and most accidents prevented. When proper safety precautions are followed, fewer accidents occur. Laboratory safety is based on three fundamental ideas:

1. Read the experiment before coming to the laboratory. Always follow the printed and verbal directions given for each experiment.

2. Use common sense when working with laboratory apparatus and materials.

3. Know how to get help in case of an accident.

A Chemistry Laboratory Safety Agreement and a Laboratory Safety Quiz are included at the end of this safety chapter. Your laboratory instructor may require you to sign the Agreement and take the Quiz.

General Safety Rules

1. **Wear splashproof goggles at all times.** Everyone in the laboratory must wear goggles that have been approved by the appropriate authorities as complying with state eye-protection laws. Previously, the American Chemical Society (ACS) recommended against wearing contact lenses in the laboratory. However, recent studies and experience have suggested that contact lenses do not increase the risk of injury to the eye. Therefore, the ACS has suspended its recommendation against wearing contact lenses in the laboratory, *provided that appropriate eye protection is also worn.* Your laboratory instructor will explain local policy regarding contact lenses in the laboratory.

2. **Know the locations and operations of all safety equipment in the laboratory.** Your laboratory instructor will show you the locations and use of such safety equipment. This equipment includes an emergency alarm or telephone system, fire extinguishers, safety or drenching showers, eyewash stations, first-aid kit, and spill kit. Do not begin work in the laboratory until you have thought out what to do in case an emergency arises.

3. **Never work alone in the laboratory.** Your laboratory instructor should be within your sight and hearing at all times, so that you can ask for help in case of questions, difficulties, or accidents.

4. **Do only the experiment assigned by your laboratory instructor.** Never substitute an unauthorized experiment for the one assigned by your laboratory

instructor. Do not deviate in any way from the designated experimental procedure without permission from your laboratory instructor.

5. **Wear clothing that is suitable for laboratory work,** including shoes made of non-porous material that completely cover the feet to protect them from spills. Tie back long hair. Wear a laboratory coat or apron, if available. Gloves may also be required for certain experiments.

6. **Immediately report all accidents to your laboratory instructor, no matter how minor.** Common laboratory accidents include chemical spills, ingestion of chemicals, cuts, burns, and fire.

7. **Never make mouth contact with any object in the laboratory.** This rule applies especially to pipets, reagents, and fingers. For this reason, no food or beverages should be present in the laboratory.

8. **Odors should be treated with caution,** because some may be toxic. Chemicals primarily enter the body through the respiratory system. Do not put your nose directly over a sample. Instead, gently waft the vapors toward your nose (see Figure 1).

Figure 1
Detecting odors

Because some toxic substances are odorless, it is vital that you be aware of the possibility of hazardous vapor. Use the fume hood when necessary. *Always* use a fume hood when handling any of the following chemicals:

bromine	carbon disulfide
chlorine	concentrated aqueous ammonia
formaldehyde	glacial acetic acid
phenol	sulfur dioxide

Highly toxic or carcinogenic substances, such as cyanides, benzene, and chloroform, must be used in a fume hood with an airflow of at least 80–120 linear feet per minute (lfm).

9. **Keep the laboratory clean and clear of unnecessary objects.** Place coats, backpacks, books, and other items unnecessary for laboratory work in the area provided for them. To prevent accidents, keep benchtops free of unnecessary

reagent bottles, glassware, and other equipment. Never place reagents or equipment on the floor.

10. **Only use equipment that is in good condition.** Chipped or cracked glassware should be repaired or discarded, as your laboratory instructor directs. Tell your laboratory instructor about any electrical equipment that is not working properly or has frayed wires.

11. **Clean up spills immediately, as directed by your laboratory instructor.** Dispose of rinses, reaction mixtures, and unused reagents as your laboratory instructor directs. Specially marked containers will be provided for discarding articles such as broken glass or ceramics, debris, paper, and burned matches. Never throw such materials into the sink or on the floor.

12. **Avoid touching hot objects.** Many chemical reactions generate heat. Therefore, do not hold a reaction vessel with your hand during a reaction. Instead, place the reaction vessel on the benchtop, or clamp it to a support stand. An apparatus that has been near a flame or other heat source may still be hot, even though it appears cool. Some materials, such as glass, retain heat for a long time, so they must be handled with extra caution.

13. **Read labels on reagent bottles and containers to make certain that they contain the appropriate reagent for the experiment.** Symbols indicating flammability, reactivity, health concerns, and other potential hazards associated with the reagent may also appear on the label. Numbers are often used to rate potential hazards on a scale from 0 (least hazardous) to 4 (most hazardous). For example, the National Fire Protection Association (NFPA) diamond label shown in Figure 2(a) indicates that the reagent is a serious hazard to your health (3), slightly flammable (1), moderately reactive (2), and has no other specific hazard associated with it. The hazard symbols found on the reagent bottle label shown in Figure 2(b) indicate that acetonitrile is flammable and that gloves should be worn while using it.

Figure 2
(a) A typical NFPA diamond label;
(b) reagent bottle label for
acetonitrile

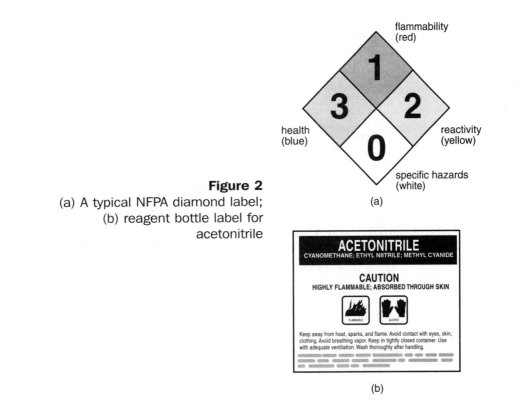

14. Use detergent or soap to wash your hands thoroughly before you leave the laboratory.

Material Safety Data Sheets (MSDS)

Additional information about hazardous reagents can be found on **Material Safety Data Sheets (MSDS).** The appropriate MSDS should be available in the laboratory where the chemical is used. Federal law requires that chemical suppliers provide appropriate MSDS to users of their chemicals.

The MSDS is divided into 16 sections, containing four types of information: (1) critical emergency information (substance identity, manufacturer, hazardous components, and identification of the hazards); (2) emergency response measures (first aid, fire-fighting and accidental release); (3) personal protection and exposure measures (handling and storage, personal protection equipment, exposure controls, physical and chemical properties, and stability and reactivity of the substance); and (4) other useful information (toxicological and ecological effects, disposal considerations, transport and regulatory information, and miscellaneous information). The layout and exact contents of MSDS vary from supplier to supplier. The first page of a typical MSDS for acetone is shown in Figure 3.

Disposal of Materials

Dispose of discarded materials as directed either by your laboratory instructor or by the instructions given in the approved experimental procedure. Assume that nothing should be discarded into the sink unless specifically directed otherwise. Soap, detergent, and water are obvious exceptions.

Never put flammable or water-immiscible liquids into the sink or any other outlet that drains into the sewage system. Such liquids include:

acetone	kerosene
diethyl ether (ethyl ether)	methanol (methyl alcohol)
ethanol (ethyl alcohol)	oils
hexane	toluene

Instead, place such liquids into appropriately labeled discard containers for disposal.

Solids should be disposed of as directed either by your laboratory instructor or by the approved experimental procedure. Place solids into appropriately labeled discard containers for subsequent disposal.

Inserting Glass into Corks or Rubber Stoppers

The following precautions must be taken when inserting glass tubing, glass rods, thermometers, funnels, or thistle tubes into a cork or a rubber stopper. First, use water or glycerol to lubricate the hole in the cork or stopper and the first 2–3 cm of the end of the piece of glass to be inserted. Drape the ends of a cloth towel across both hands. Hold the cork or stopper in one hand. With the other hand, grasp the glass tubing close to the end you wish to insert (see Figure 4 on page 6).

Carefully insert the lubricated end of the tubing into the stopper, using gentle pressure and a twisting motion. Slowly work the tubing through the stopper.

Section 1   Chemical Product and Company Identification

MSDS name: acetone

Catalog number(s): AB89012, BC45721

Manufacturer/Supplier: ABC Supply

Information: 555/555-1003

Emergency contact: In the event of a spill, leak, fire, explosion, or accident, call 555/555-1004

Section 2   Composition and Information on Ingredients

CAS No. 67-64-1, EC No.: 200-662-2

Synonyms: aceton, Chevron acetone, dimethylformaldehyde, dimethylketal, dimethyl ketone, ketone propane, *beta*-ketopropane, methyl ketone, propanone, 2-propanone, pyroacetic acid, pyroacetic ether

RCRA waste number: U002

Section 3   Hazards Identification

NFPA rating (scale 0–4)

    HEALTH 1

    FIRE 3

    REACTIVITY 0

    Emergency Overview: Label FLAMMABLE. Keep away from ignition sources. No smoking. Keep container tightly closed and in a cool place.

    Potential health effects: Irritating to respiratory system and skin. Risk of serious damage to eyes. Target organs include liver and kidneys. In case of contact with eyes, rinse immediately with plenty of water, and seek medical advice. Wear suitable protective clothing.

Section 4   First Aid Measures

    If swallowed, wash out mouth with water, provided the person is conscious. Call a physician. If inhaled, remove person to fresh air. If person is not breathing, give artificial respiration. If person's breathing is difficult, give oxygen. In case of skin contact, immediately wash skin with soap, rinsing with a large amount of running water. In case of eye contact, immediately flush eyes with a large amount of running water for at least 15 min.

Section 5   Fire-Fighting Measures

    Extinguishing media: Water spray, carbon dioxide, dry chemical powder, or appropriate foam. Note that water will be effective for cooling, but may not extinguish fire.

    Special fire-fighting procedures: Wear self-contained breathing apparatus and protective clothing in order to prevent contact with skin and eyes.

Section 6   Accidental Release Measures

    Evacuate area. Wear respirator, chemical safety goggles, rubber boots, and rubber gloves. Cover spill with dry lime, sand, or soda ash. Shut off all sources of ignition. Place in covered containers, using non-sparking tools, and trasport outdoors. Ventilate area, and wash spill site after material pickup is complete.

Section 7   Handling and Storage (refer to Section 8)

Section 8   Exposure Controls and Personal Protection

    Safety shower and eye bath: Use non-sparking tools. Mechanical exhaust required. Wash thoroughly after handling. Wash contaminated clothing before reuse. Avoid breathing vapor. Avoid contact with eyes, skin, and clothing. Avoid prolonged or repeated exposure. NIOSH/MSHA-approved respirator. Compatible chemical-resistant gloves. Chemical safety goggles. Keep container closed. Keep away from heat, sparks, and open flame. Store in a cool, dry place. The material may slowly penetrate protective gloves; therefore, in case of spills, discard gloves after use.

Section 9   Physical and Chemical Properties

Appearance and odor: liquid, odorless	Explosion limits in air: upper, 13%; lower, 2%
Boiling point: 56 °C	Vapor pressure: 400 mm Hg
Melting point: –94 °C	Specific gravity: 0.791
Flash point: 1 °F/–17 °C	Vapor density: 2 g/L

Figure 3   Typical MSDS information for acetone

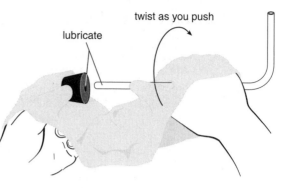

Figure 4
Inserting glass tubing into a stopper

As soon as enough tubing extends through the stopper to grasp with your fingers, pull the tubing through the stopper to the desired length.

When inserting a bent piece of glass tubing into a stopper, do not exert pressure on the bend. Instead, grasp the tubing as close as possible to the end being inserted, and apply pressure there.

Fire Safety

If a fire occurs accidentally in the laboratory, it is important to stay calm. ***In most cases, it is best to evacuate the laboratory and let the laboratory instructor deal with the fire.***

Inhalation of smoke and toxic fumes can be as dangerous as burns caused by flames. If a large fire breaks out, the building should be evacuated. The fire department should be called immediately and told the exact location and cause of the fire.

When used appropriately, fire extinguishers are the best devices for putting out a ***small*** fire. Different types of extinguishers should be used on different kinds of fires. Your laboratory instructor will describe the locations and uses of the fire extinguishers in your laboratory. However, remember that it is usually best to evacuate the laboratory and get help from someone trained in the use of fire-fighting equipment.

In addition to using fire extinguishers, there are other ways to put out ***small*** fires. One is to remove the heat source, if possible, then place an inverted beaker or similar item over the fire to cut off the supply of oxygen to the fire. Putting water-soaked towels on the burning area will also work.

If your clothing starts burning, move away from the heat source. ***Call for help. STOP*** what you are doing. ***DROP*** to the floor. ***ROLL*** over repeatedly to try to extinguish the flames until someone can use a fire blanket to smother them. ***Do not run*** to the fire blanket or safety shower. The fire blanket should be removed as soon as the flames are extinguished. Your laboratory instructor may place you under the safety shower at this point.

While a burn victim is being helped, others should try to shut off or reduce the fuel supply to the source of the fire. If possible, direct the spray from an appropriate fire extinguisher at the ***base*** of the fire. Alert appropriate authorities about the fire and its status. Call for additional assistance if needed.

▌▌ Glassware and General Equipment

Laboratory apparatus used in general chemistry can be divided into two broad categories: glassware and equipment. Each piece of apparatus is designed to be used for one or more specific purposes.

Figures 5(a) and 5(b) on pages 10 and 11 show illustrations, names, uses, and sizes of most glassware and general equipment. Be sure you know the names and uses of each item shown.

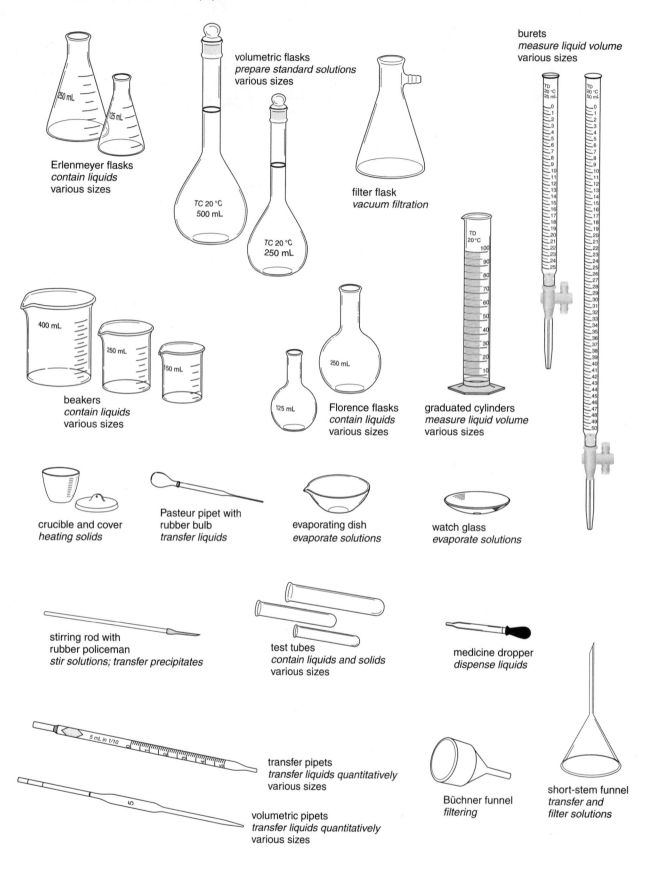

Figure 5(a)   Illustrations of glassware, including names, uses, and sizes

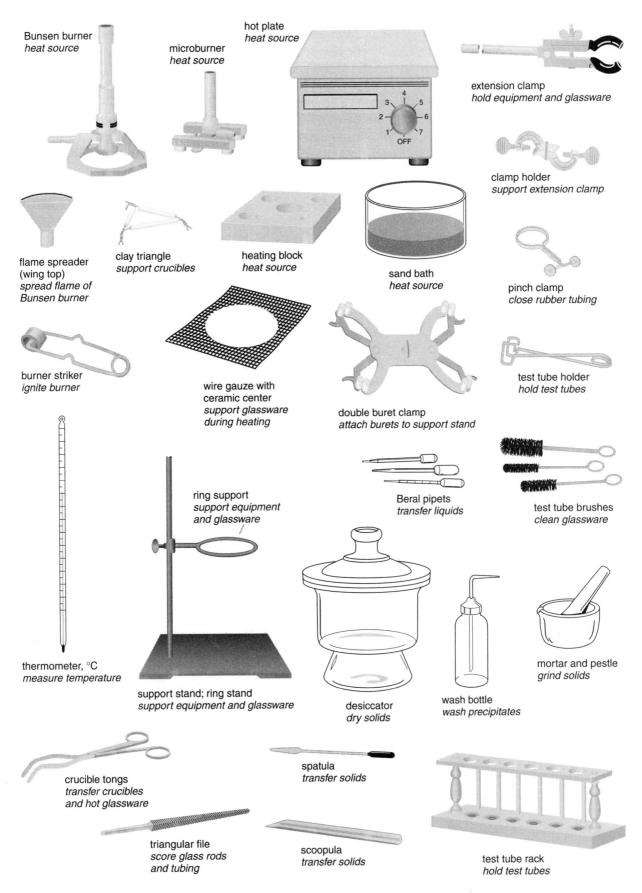

Figure 5(b)   Illustrations of equipment, including names and uses

© 2002 Wadsworth Group

Weighing

Solid and liquid masses are measured using **balances.** Three types of balances are discussed in this chapter: top-loading, triple-beam, and analytical. Your laboratory instructor will give you specific directions for using the balances in your laboratory.

General rules for balance use include the following:

1. *Never* place chemicals directly on a balance pan. Instead, place chemicals in a pre-weighed container or on a pre-weighed piece of weighing paper.

2. Use tongs or a clean piece of laboratory tissue to handle the container or weighing paper holding the sample being weighed.

3. Do not place hot objects on a balance pan. The material being weighed must be at the same temperature as the balance to ensure accurate mass measurement.

4. Inform your laboratory instructor if the balance is not level or at the zero mark when empty. Do not attempt to level the balance or adjust the zero mark yourself.

5. Always make sure that the balance is at the zero mark before you begin weighing **and after you finish weighing.** If the balance does not register zero at these times, consult your laboratory instructor.

6. Immediately clean up any chemicals spilled on or near a balance, following the directions of your laboratory instructor.

Using a Top-Loading Balance

Top-loading balances (see Figure 6) are used for rapid determination of masses to the nearest 0.1–0.001 g (depending on the model). Turn on a top-loading balance by pressing the control bar or button. The display should read "0.000 g". If it does not, inform your laboratory instructor.

Using tongs or tissue, place the empty sample container on the center of the pan. Press the control bar, or tare bar (if there is one), to display the container mass. Press the tare bar a second time to return the display to zero. If the balance does not have a tare bar, record the mass of the empty container at this point.

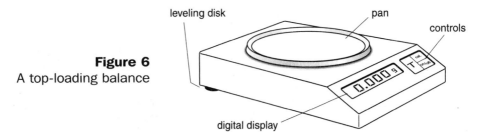

Figure 6
A top-loading balance

Once an empty container has been weighed using the tare function, the display is automatically reset to zero whenever that empty container is on the pan. The **tare** is the mass of the empty container, which is retained in the balance's memory and automatically subtracted from the total mass of the container and its contents. Therefore, the display will show only the mass of the contents.

Remove the empty container from the balance. Place the substance or object to be weighed in the container. Replace the container on the center of the balance pan. If the balance has a tare function, the display will show only the mass of the substance or object. If not you must subtract the mass of the empty container from the mass shown on the display to obtain the mass of the sample.

Read and record the sample's mass, using the number of figures to the right of the decimal point specified by your laboratory instructor. Using tongs or tissue, remove the container and its contents from the balance pan and turn off the balance.

Using a Triple-Beam Balance

Triple-beam balances have three beams with sliding masses, have about 610-g capacity, and are used to weigh to the nearest 0.1 g (see Figure 7). The front beam is graduated to 10 g in 0.1-g divisions, the center beam to 500 g in 100-g divisions, and the rear beam to 100 g in 10-g divisions.

The mass of a sample is determined by moving the sliding masses on the beams until the sliding masses exactly balance the sample. With all three sliding masses at their zero positions, placing a sample on the pan causes the pointer to deflect upward from the zero mark. This movement indicates that the sliding masses are not far enough to the right on the beams to balance the sample. However, when the sliding masses are moved too far to the right, they overbalance the sample in the pan, and the pointer deflects downward. The sliding masses exactly balance the sample when the pointer swings equal distances above and below the zero mark. At this point you do not need to wait until the pointer stops moving to determine that the sliding masses and sample balance.

Begin by placing the three sliding masses at zero. Check to see that the pointer freely swings an equal distance above and below the zero mark. Consult your laboratory instructor if this is not the case.

Use crucible tongs or laboratory tissue to place the object to be weighed on the center of the pan. Move the sliding mass on the center beam to the 100-g notch. If the pointer still deflects upward, move the middle sliding mass to the 200-g notch. Continue moving the middle sliding mass to the right until the pointer deflects downward. Then move the mass one notch to the left.

Use the same procedure to position the 10-g and 1-g sliding masses so that the three masses exactly balance the mass of the object on the pan [see Figure 7(b)]. The mass of the sample is the sum of the masses indicated on the three beams minus the mass of the sample container. Record the sample mass using the number of significant figures specified by your laboratory instructor. Remove the sample container from the balance, and reset the sliding masses to their zero positions.

There are also **four-beam balances,** capable of measuring to the nearest 0.01 g [Figure 7(c)]. Three of the beams are identical to those on a triple-beam balance. The fourth beam is graduated to 1.00 g in 0.01-g divisions. Except for the fourth beam, these balances are operated like a three-beam balance.

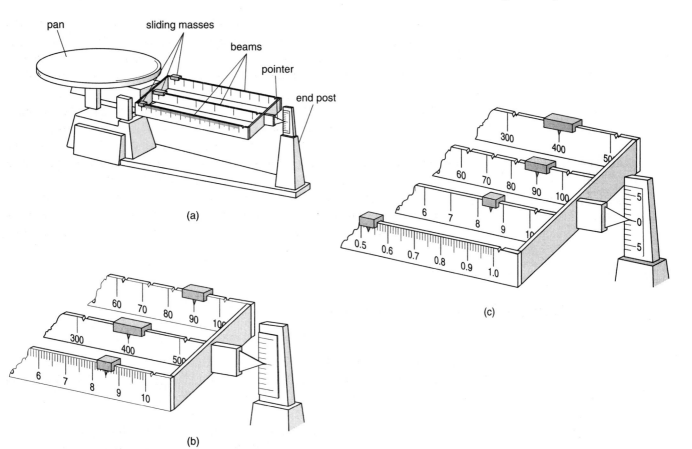

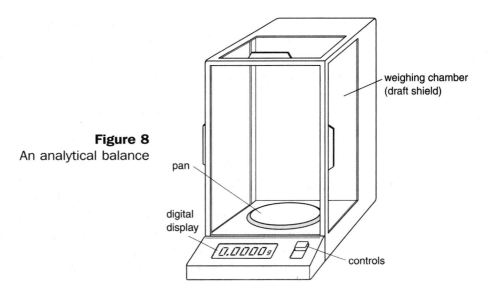

Figure 7 (a) A triple-beam balance; (b) reading mass on a triple-beam balance (498.5 g shown); (c) a four-beam balance

Using an Analytical Balance

Analytical balances (Figure 8) are used for very accurate, quantitative mass measurements to the nearest 0.0001 g. These balances are much more delicate than either top-loading or triple-beam balances. Therefore, the general rules for using balances listed at the beginning of this chapter *must* be followed scrupulously to avoid damaging the balance.

Figure 8
An analytical balance

An analytical balance looks like a top-loading balance with a glass-sided box surrounding the pan. This box, or **weighing chamber,** acts as a draft shield, so that the very accurate weighings will not be affected by any air currents in the laboratory. The weighing chamber has doors that allow access to the pan from the sides and the top.

Although the controls may vary a bit from model to model, the following general procedure applies to all analytical balances.

Begin by checking to see that the balance is level. If it is not level, inform your laboratory instructor who will adjust the leveling disks normally located on two feet of the balance. Also check to see that the balance pan is clean and that the weighing-chamber doors are closed.

Press the on/off button or control bar to turn on the balance. The display should indicate 0.0000 g. If it does not, ask your laboratory instructor for assistance.

To weigh a solid object that is **_not_** a reagent, open one of the weighing-chamber doors. Gently place the solid on the center of the pan. Close the door. The mass of the object will appear on the display. Record the mass.

To tare a container and then determine the mass of reagent added to it, open a weighing-chamber door and gently place the empty container on the center of the pan. Close the door and press the button or control bar that turns on the balance. The mass of the container will appear on the display. Record the mass of the container. Press the tare button or control bar. The tare function will reset the display to 0.0000 g with the container on the pan. From this point on, the display will show only the mass of reagent added to the container.

Next open a weighing-chamber door. Slowly and carefully add reagent to the tared container until the desired amount has been added. The display will indicate your progress. The reading may fluctuate for a few seconds after you stop adding reagent, but will stabilize in a short time. When you have finished adding reagent, close the weighing-chamber door. Record the final mass of the reagent from the display.

Press the on/off button or control bar to turn off the balance. Remove the container from the pan, and close the weighing-chamber door.

Questions Regarding Weighing

1. A student needs to determine the mass of a liquid to the nearest 0.1 g. What type of balance could be used? If the mass had to be determined to the nearest 0.01 g, what kind of balance should be used?

2. Explain what the tare function does on an electronic balance.

3. A student used a triple-beam balance to determine the mass of a solid and reported the mass to be 21.347 g. Is this reported mass experimentally valid? Explain.

4. A student recorded the mass of a two-component solid mixture, then separated the two components and determined their individual masses. The student found the mass of one of the components to be greater than the mass of the original mixture. Explain how such a result could occur.

5. Explain why a piece of laboratory tissue or tongs should be used to handle a sample container to be weighed.

6. When using a triple-beam balance, you don't need to wait until the pointer stops moving. What indicates that the correct mass has been read?

IV

Measuring Liquid Volumes

Volumetric glassware, unlike other glassware, has calibration marks used to determine the specific volume of liquid contained in the glassware. Some volumetric glassware, such as graduated pipets and burets, is used to accurately dispense measured volumes of liquids. Other volumetric glassware, such as graduated cylinders, volumetric pipets, and volumetric flasks, is used to contain exact volumes of solutions.

A liquid contained in plastic tube has a flat surface, so the liquid's volume can be read easily. However, a liquid in a glass tube has a curved surface known as a **meniscus.** For most liquids, the meniscus is concave, and the bottom of the curve is used to determine the volume of the liquid (see Figure 9).

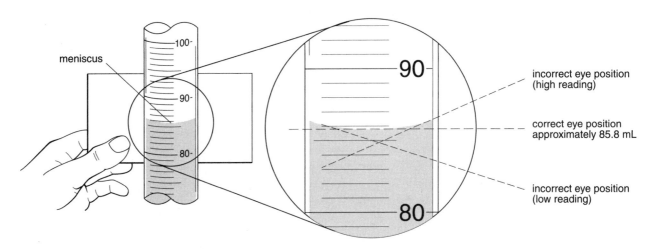

Figure 9   Measuring liquid volume in a glass graduated cylinder

When reading the volume of a liquid in a piece of volumetric glassware, first note the scale marked on the glassware and the position of the meniscus. Position your eye so it is level with the bottom of the meniscus. Look straight at the meniscus through the glassware so that you see only a concave line. Read the level of the bottom of the meniscus, as shown in Figure 9.

If your eye is above the meniscus, you will obtain an incorrectly high reading. If your eye is below the meniscus, you will obtain an incorrectly low reading. This apparent variation in volume, due to viewing the meniscus from different angles, is known as **parallax.**

Holding a white card or paper directly behind the glassware at the meniscus level can help make liquid volume measurement more accurate. Proper reading of the meniscus and avoiding error due to parallax are vital when using volumetric glassware.

Graduated Cylinders

Graduated cylinders (see Figure 9) are made of either plastic or glass. They vary in capacity from 5 mL to 2000 mL. The cylinders are usually marked in units that are about 1% of the cylinder's total capacity. Most cylinders are calibrated to deliver (TD) the specified volume and are therefore equipped with a pouring spout.

Pipets

Two basic types of pipets are used to quantitatively measure liquid volume. The first type is the **graduated,** or **measuring, pipet,** calibrated in convenient units to deliver volumes up to the pipet's maximum capacity. These pipets are drained from a higher calibration mark to a lower mark on the pipet in order to deliver a specific volume.

The second type is the **volumetric,** or **transfer, pipet** which is used to deliver a single, fixed volume of liquid. Volumetric pipets are calibrated with a single mark (indicating their maximum capacity). They are available in capacities ranging from 1 mL to 200 mL.

Using Volumetric Pipets

These instructions are for using volumetric pipets. With minor modifications, they can also be used for graduated pipets. The instructions describe the proper ways to obtain a solution from a stock supply, prepare a pipet, and use a pipet to transfer a measured volume of solution. Instructions for cleaning pipets (and other glassware) are given in Appendix A: Operations with Glass.

Solutions and liquids are drawn into a pipet by applying a slight vacuum at the top of the pipet, using a rubber suction bulb or other pipet-filling device. Your laboratory instructor will demonstrate the use of a rubber bulb or other device to fill a pipet. *Never use your mouth to pipet any liquid or solution.*

First, pour a small amount of the solution to be pipetted into a clean, dry beaker. If you are using a rubber suction bulb, partially squeeze the bulb and slip it onto the wider end of a clean pipet, as shown in Figure 10(a). Make certain that the connection is secure enough to prevent air leaks. If you are using another type of pipet-filling device, follow your laboratory instructor's directions.

Hold the pipet tip below the solution surface in the beaker, as shown in Figure 10(b). Keep the tip well below the surface throughout the pipetting so that air is not drawn into the pipet along with the solution. Draw a small portion of solution into the pipet by slowly releasing pressure on the bulb. Be careful *not* to draw liquid into the bulb. This small initial amount of solution is used to rinse the pipet to avoid diluting the pipetted sample with any distilled water remaining in the pipet after cleaning.

Quickly but gently disconnect the bulb. Immediately place your index finger on top of the pipet to prevent the solution from draining out. Hold the pipet nearly horizontal, and rotate it to allow the solution to contact all interior surfaces. Lift your index finger to allow the solution to drain through the pipet tip into a discard container. Discard this solution as directed by your laboratory instructor. Repeat this rinsing step at least twice with small portions of the solution.

The next step uses the rinsed pipet to transfer a measured volume of the solution from the beaker to another container. Use a bulb or other device to draw enough solution into the pipet so that the liquid level is *above* the calibration mark.

Lightly rest the tip of the pipet against the bottom of the beaker. Gently remove the bulb [see Figure 10(b)] and quickly press your index finger on top of the pipet, as before. ***Do not allow the liquid level to fall below the calibration mark.*** A ***slightly*** moistened finger on top of the pipet provides the best flow control.

Remove the pipet from the solution, keeping your index finger on top. Use a clean tissue to wipe off any liquid on the outside of the pipet. Slowly and carefully lift your index finger to drain some of the solution into the beaker until the bottom of the meniscus aligns exactly with the calibration mark [see Figure 10(c)]. When the bottom of the meniscus is at the calibration mark, quickly press your index finger on top of the pipet to stop the liquid flow. Touch the tip of the pipet against the inner wall of the beaker so that the hanging drop is transferred to the beaker [see Figure 10(c)].

Move the pipet to the container into which you wish to transfer the measured solution. Hold the tip of the pipet against the inner wall of the container [see Figure 10(d)]. Lift your index finger from the pipet and allow the solution to drain down the inner wall of the container. When the flow stops, hold the pipet vertically for 15 seconds more to allow for complete draining. ***Do not attempt to remove the small amount of solution remaining in the tip of the volumetric pipet.*** The pipet was calibrated to transfer an exact volume of liquid ***excluding*** the liquid remaining in its tip [see Figure 10(e)]. Touch the pipet tip to the inner wall of the container to transfer the hanging drop to the container.

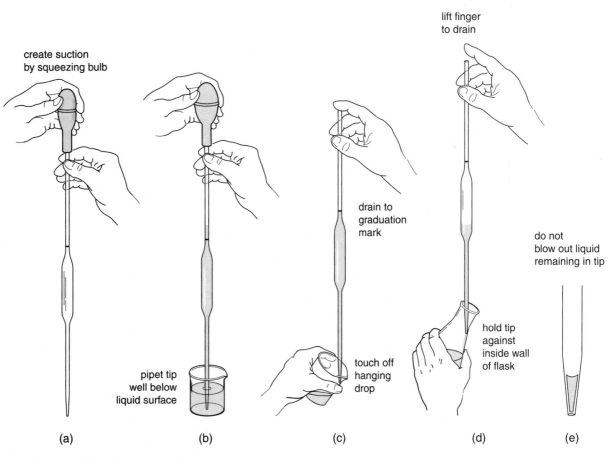

create suction by squeezing bulb

pipet tip well below liquid surface

drain to graduation mark

touch off hanging drop

lift finger to drain

hold tip against inside wall of flask

do not blow out liquid remaining in tip

(a)   (b)   (c)   (d)   (e)

Figure 10 Using a volumetric pipet

In contrast, when liquids are being transferred using a **graduated pipet,** the liquid level should not be allowed to fall below the lowest calibration mark into the uncalibrated section of the pipet.

Burets

A **buret** is a long, narrow, calibrated tube with a device at one end to control the flow of liquid from the buret. The most common buret sizes are 10 mL, 25 mL, and 50 mL. Measurements made using burets are more precise than those made using a measuring pipet. Burets differ chiefly in the type of flow-control device, or valve, that is used. Two common buret valves are: a tapered glass stopcock at the end of a glass barrel, and a Teflon stopcock attached to a glass or Teflon barrel (see Figure 11).

Figure 11
Two common buret types:
(a) glass barrel with tapered glass stopcock; (b) Teflon barrel with Teflon stopcock

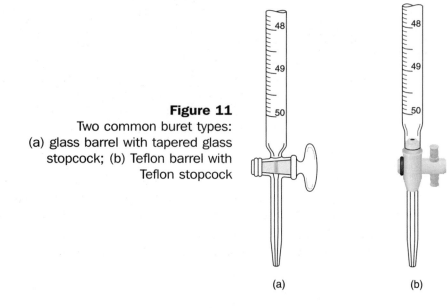

(a)   (b)

Teflon stopcocks do not need lubrication, but glass stopcocks require a thin film of stopcock grease. To lubricate a glass stopcock, carefully remove the stopcock from the barrel. Wipe the old grease off the stopcock and the barrel. Spread a small amount of stopcock grease thinly over the stopcock surface, avoiding the area near the hole. Insert the stopcock into the barrel. Rotate the stopcock several times in one direction. The stopcock should now appear nearly transparent where it contacts the barrel.

Before using a buret, check to see that the stopcock is liquid tight by closing the stopcock and filling the buret with water. Let the buret stand for a few minutes, then check for leaks. Consult your laboratory instructor if your stopcock leaks.

A buret must be clean if it is to yield accurate measurements. See Appendix A for instructions about cleaning burets.

When the buret is clean and ready to use, rinse it three or more times with small portions of the solution to be used in the buret. Dispose of the rinses as specified by your laboratory instructor. Clamp the buret vertically to a ring stand. Close the stopcock and use a short-stem funnel to slowly fill the buret with the sample solution. Do not allow the solution to overflow the funnel. Fill the buret to above the zero calibration mark at the top. Remove the funnel.

Place a discard beaker under the stopcock, open the stopcock, and drain solution until the meniscus is at, or slightly below, the zero calibration mark. To remove the hanging drop, touch the wet inside wall of the beaker to the buret tip.

It wastes time to exactly align the meniscus with the zero mark. Simply read the meniscus wherever it is within the calibrated portion of the buret, and record this reading as the initial volume. Read the meniscus in a buret with great care. A 50-mL buret is calibrated in units of 0.1 mL, but one-fifth of a division can be reproducibly estimated. Therefore, estimate the position of the meniscus if it is between calibration marks, and record buret readings to the nearest 0.02 mL. Holding a white index card or paper directly behind the meniscus in the glassware can improve the accuracy of volume measurements (see Figure 12).

Figure 12

Reading the meniscus in a buret

reading is
32.44 mL

When using a buret with a Teflon stopcock, operate the stopcock with either hand without the risk of the stopcock slipping out. Glass stopcocks should be operated with the hand on the opposite side of the buret barrel from the stopcock, to avoid pulling the stopcock from the barrel (see Figure 13). Light pressure from the hand turning the stopcock will prevent it from slipping out.

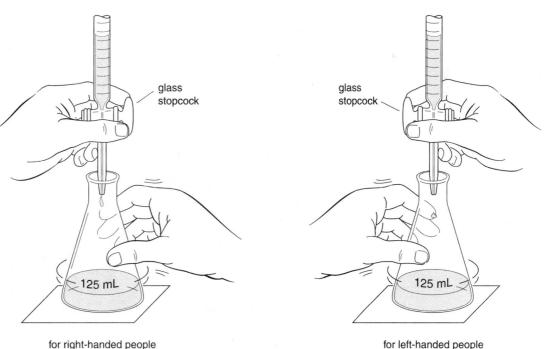

glass
stopcock

glass
stopcock

125 mL

125 mL

for right-handed people

for left-handed people

Figure 13 Using a buret

Drain solution from a buret relatively slowly, so that the solution drains reproducibly. After you have drained the desired volume, close the stopcock. Touch the inner wall of the flask to the buret tip to remove the hanging drop. Use distilled water from a wash bottle to rinse the drop into the liquid in the flask. Record the final buret reading.

When finished with the buret, drain any remaining solution, and dispose of the solution as specified by your laboratory instructor. Clean the buret thoroughly (see Appendix A: Cleaning Glassware).

Volumetric Flasks

Volumetric flasks, available in 5-mL to 5-L capacities, are calibrated to contain a specified volume (see Figure 14). The single calibration mark is located on the narrow neck to allow filling to a reproducible volume.

Volumetric flasks are used to prepare solutions in two general ways. The first involves dissolving a solid solute to make a solution of a definite volume. To do so, weigh the solute, then dissolve it in the minimum amount of solvent in a clean beaker. Using a funnel, transfer the solution to a volumetric flask. Rinse the beaker with a small amount of the same solvent, and transfer the rinse through the funnel into the flask. Repeat this rinsing procedure several times to thoroughly rinse the beaker and the funnel.

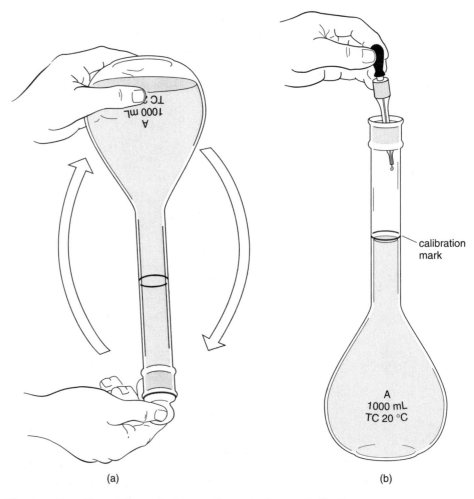

(a)   (b)

Figure 14 Preparing solutions using a volumetric flask

The second way to prepare a solution using a volumetric flask involves the quantitative dilution of a concentrated solution. Begin by using a pipet to transfer an accurately measured sample of the concentrated solution to the volumetric flask. Techniques for using pipets are given on page 22.

With either procedure, next add more of the same solvent to the flask until it is about two-thirds full, while slowly swirling the flask to agitate the solution. Continue slowly adding more solvent until the liquid level is just below the calibration mark. Stopper the flask, and thoroughly mix the solution by inverting the flask end over end ten times [see Figure 14(a)]. Use a small dropper or a Pasteur pipet to add the final amount of solvent needed to align the bottom of the meniscus with the calibration mark [see Figure 14(b)]. Avoid parallax error when reading the final liquid level.

Once the bottom of the meniscus is aligned with the calibration mark, firmly seal the flask using a stopper or plastic cap. Hold the stopper or cap firmly in place and slowly invert the flask 10 to 15 times to make a homogeneous solution.

Solutions are usually not stored in volumetric flasks but transferred to clean, dry, labeled bottles for storage. After transferring the solution, thoroughly rinse the volumetric flask and invert it to dry. Store the dry flask with the stopper or cap in place. If the stopper is glass, insert a small strip of paper between the stopper and the flask to prevent sticking.

Questions Regarding Measuring Liquid Volumes

1. A student needs to measure 4.37 mL of a liquid. Which of these pieces of glassware could be used? (a) 25-mL graduated cylinder; (b) 10-mL graduated cylinder; (c) unmarked Beral pipet; (d) transfer pipet; (e) measuring pipet. Explain your answer.

2. Two students read the volume of liquid in a buret. The first student reads the volume as 31.66 mL. The second student reads the volume as 31.70 mL, which is the correct value. What is a likely explanation for the difference between the volume readings?

3. When liquid is drawn into a volumetric pipet, why should the liquid level not be allowed to fall below the calibration mark?

4. A student making up a solution in a volumetric flask adds solvent to just above the calibration mark. Will the concentration of the resulting solution be higher or lower than that desired?

5. Which pieces of laboratory equipment shown in Figure 5(a) and (b) would be required to determine the density of a liquid? Name any other pieces of equipment not shown in the figures that would also be needed.

6. Which pieces of laboratory equipment shown in Figure 5(a) and (b) would be needed to measure and transfer 5.6 mL of an organic liquid to a beaker, then heated to 47.0 °C?

V
Transferring Materials

When transferring a reagent from one container to another, read the label on the reagent container **twice before** removing any reagent, to make sure that it is the appropriate material. After completing the transfer, read the label a **third** time to confirm that you transferred the correct material.

Transferring Solids

Solids are somewhat more difficult to transfer than are liquids.

Take a labeled, wide-mouth container, such as a beaker, to the reagent shelf to obtain the solid required. During transfer of a solid, either hold the reagent bottle cover in your fingers, or lay it on the benchtop with the top side down to prevent contamination, as shown in Figure 15(a).

Pour solids into your container by gently tipping the reagent bottle while slowly rotating it back and forth, as shown in Figure 15(a). Just tipping the bottle may cause large lumps of the reagent to drop suddenly into your container possibly causing a spill.

Use a spatula only to transfer solids from **your** container, as shown in Figure 15(b). Do **not** put your spatula into a reagent bottle.

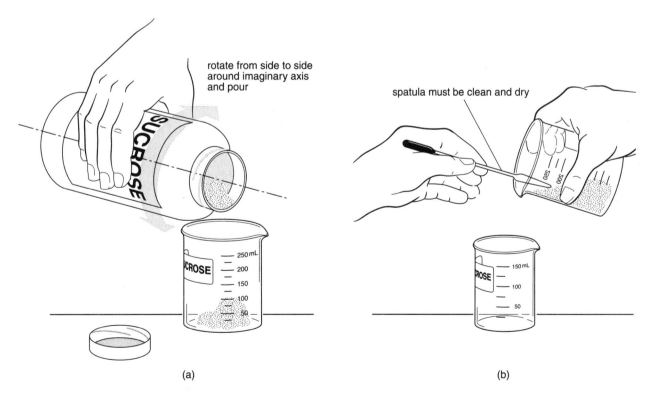

(a)

rotate from side to side around imaginary axis and pour

spatula must be clean and dry

(b)

Figure 15 Transferring a solid from (a) a reagent bottle and (b) from a smaller container, using a spatula

Replace the original cover on the reagent bottle. Interchanging reagent bottle covers can cause contamination. Do *not* return unused solids to their reagent bottles; this can also cause contamination.

Transferring Liquids

Take an appropriately labeled container to the reagent shelf. Transfer the required liquid from the reagent bottle directly into your container over a nearby sink to catch any spills.

Transferring from a stoppered bottle: Pour the liquid from the reagent bottle into your labeled container. During the transfer, either hold the stopper in your fingers or place it *upside down* on a flat surface. To avoid contamination, do *not* lay a glass stopper on its side. Proper technique is shown in Figure 16(a). Be sure the original stopper is returned to the reagent bottle. Do *not* interchange stoppers.

Transferring from a dropper bottle: Take care not to contaminate the contents of the dropper bottle. *Never* lay the dropper on any surface. Be sure the dropper *never* touches your container or its contents. Proper transfer of a liquid using a dropper is shown in Figure 16(b).

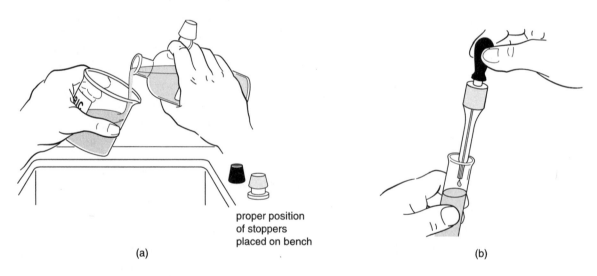

proper position
of stoppers
placed on bench

(a)   (b)

Figure 16 Proper technique for transferring a liquid from (a) a stoppered reagent bottle and (b) a dropper bottle

If you use a Beral or Pasteur pipet to transfer a liquid, do *not* pipet directly from the reagent bottle unless instructed to do so. Instead, pour some of the liquid into your labeled container, and then pipet from your container.

Using Pipets Non-Quantitatively

Beral and Pasteur pipets are commonly used for non-quantitative transfer of small quantities of liquids. For example, they can be used either to add liquid to a container or to draw off liquid from a mixture.

Mixing Liquids

To dilute concentrated liquids with water, ***always*** add the concentrated reagent to the water. This ensures that the resulting solution will be dilute. Slowly add small amounts of the concentrated reagent. It is especially important to follow this procedure when adding a concentrated acid to water, because of the large amount of heat generated by this combination. If any splattering occurs, it is more likely to be of the dilute solution, which is not as harmful as splattering of the concentrated acid.

Mixing a Solid and a Liquid

When mixing a solid and a liquid, add the solid to the liquid, while stirring continuously. Add the solid in small amounts, unless instructed otherwise.

Questions Regarding Transferring Materials

1. A student returned some unused solid to its reagent bottle. Why was this improper procedure?

2. Explain why you should read the label on a reagent container several times when handling the reagent.

3. Explain why it is best not to pipet a liquid directly from a reagent bottle.

4. A student in a hurry decided to make up a dilute sulfuric acid solution by adding water to the concentrated acid. What likely occurred?

5. Explain why a dropper from a dropper bottle should never be placed on a benchtop.

6. A student used a spatula to transfer a solid directly from a reagent bottle into a test tube containing a liquid. What was wrong with this procedure?

VI

Heating Sources and Techniques

Heat is frequently used to increase reaction rates, enhance evaporation, or speed dissolution. Gas burners, usually Bunsen burners, and hot plates are the two most common heat sources.

Bunsen Burner

The **Bunsen burner** was invented in 1855 by Robert Bunsen, a German chemist. A typical Bunsen burner is shown in Figure 17(a). If your gas burner differs from the one described here, your laboratory instructor will give you directions for its use.

Combustible gas flows through rubber or plastic tubing from a gas cock on the laboratory bench to the gas inlet of the burner. A typical Bunsen burner has a metal mixing tube, or barrel, attached to its base, as shown in Figure 17. Gas passing up the burner barrel mixes with air drawn into the barrel through the air inlets. The gas–air mixture is ignited by holding a burning match just above the top of the barrel, as shown in Figure 17(b). Alternatively, an inverted striker is held just above the barrel top, and the striker handle squeezed to create a spark.

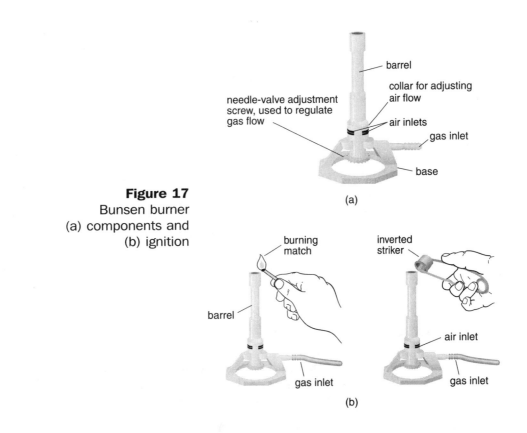

Figure 17
Bunsen burner
(a) components and
(b) ignition

The type of flame is controlled by varying the relative amounts of gas and air entering the barrel. Major adjustments in gas flow are made by turning the handle on the bench gas cock. More precise control is achieved by careful adjustment of the needle-valve screw at the base of the burner [see Figure 17(a)]. The amount of air entering the barrel is controlled by rotating the barrel to reposition the collar. The rotation causes the collar to cover the air inlet holes to a greater or lesser extent [see Figure 17(a)].

If the flame goes out, ***turn off the gas, and wait at least 30 seconds*** before attempting to relight the burner to allow the uncombusted gas to disperse.

Different flame temperatures result from different gas–air mixtures. A mixture that is mostly gas with only a little air produces a relatively cool yellow flame, referred to as a ***reducing,*** or luminous, flame [see Figure 18(a)]. As more air is mixed with the gas, the temperature of the flame increases. Efficiently burned gas produces a hot flame, called an ***oxidizing,*** or nonluminous, flame. An oxidizing flame has an inner flame that is bright blue and an outer flame that is almost colorless [see Figure 18(b)]. The hottest portion of an oxidizing flame is just above the peak of the blue inner flame.

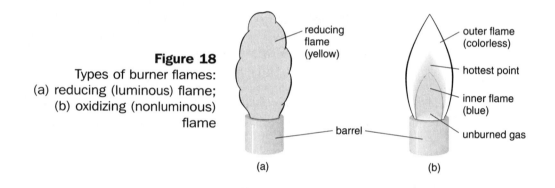

Figure 18
Types of burner flames:
(a) reducing (luminous) flame;
(b) oxidizing (nonluminous) flame

reducing flame (yellow)

outer flame (colorless)

hottest point

inner flame (blue)

barrel

unburned gas

(a)   (b)

If the gas begins to burn inside the barrel, immediately turn off the gas cock. Allow the barrel to cool. Then adjust the collar to decrease the amount of air entering the barrel (see Figure 17). Relight the burner, and make further adjustments to obtain the proper mixture of gas and air.

Microburner

A **microburner** is a smaller version of a Bunsen burner, as shown in Figure 19. The principal difference between the two types of burners, other than size, is the way the gas–air mixture is controlled. In a microburner, gas flow is regulated at the gas cock only, and air intake is controlled by adjusting the position of a flat metal plate at the base of the burner. The plate limits the amount of air that mixes with the gas in the barrel. A microburner is ignited in the same manner as the Bunsen burner.

Hot Plate

A **hot plate** is an electrical heating device that is particularly useful for experiments where an open flame is prohibited. A dial or knob controls the hot plate temperature (see Figure 20).

© 2002 Wadsworth Group

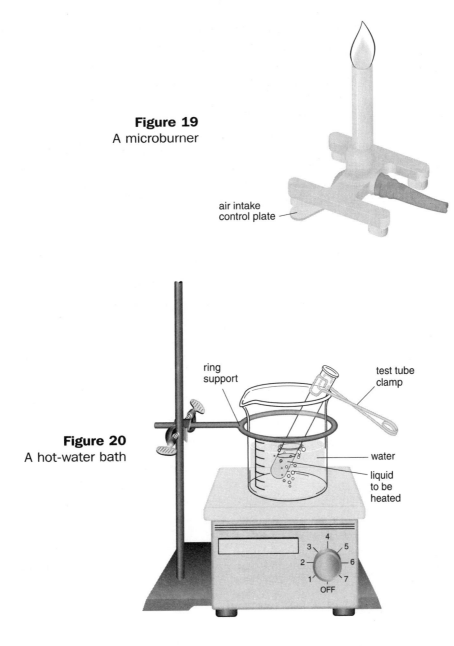

Figure 19
A microburner

air intake
control plate

Figure 20
A hot-water bath

ring
support

test tube
clamp

water

liquid
to be
heated

4
3   5
2   6
1   7
OFF

Heating a Liquid

Use caution when heating liquids, especially flammable liquids or small volumes of *any* liquid. Heat flammable liquids only with a hot-water bath or on a hot plate especially designed for heating flammable liquids, under the careful supervision of your laboratory instructor.

A hot-water bath can also be used to heat a small amount of liquid in a test tube to temperatures of up to 100 °C. To assemble a hot-water bath, half fill a beaker with water. The beaker should be about three-quarters as tall as the test tube being heated (see Figure 20). Place the beaker on a hot plate or on a ceramic-centered wire gauze, supported by a ring stand, over a Bunsen burner. To prevent the beaker from being tipped accidentally, surround it with a support ring attached to a ring stand, as shown in Figure 20. Heat the water to the desired temperature, which is not necessarily boiling. Then half fill a test tube with the liquid to be heated. Using a test tube holder, position the tube so that all of the sample liquid is below the surface of the water in the bath (see Figure 20). Heat the sample until it reaches the desired temperature.

If a liquid in a test tube must be heated above 100 °C, the test tube can be heated directly with a burner flame, but cautiously. The test tube must be no more than one-third full of liquid. Use a test tube holder to hold the tube at a 45° angle. Gently heat the test tube contents by positioning the tube so that the tip of the burner flame is just below the surface of the liquid in the tube (see Figure 21). Move the tube in and out of the flame to control the heating rate, as shown in Figure 21. Do not directly heat the bottom of the tube. Improper heating can cause hot liquid to suddenly spurt from the tube. Therefore, when heating a sample in a test tube, ***never*** look into the tube or point the open end toward anyone.

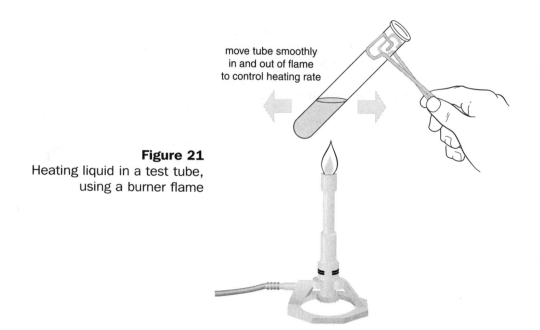

move tube smoothly
in and out of flame
to control heating rate

Figure 21
Heating liquid in a test tube,
using a burner flame

Heating a Solid

Porcelain crucibles are commonly used for heating solids to thermally decompose the solids or to heat them to dryness. Usually, the empty crucible is first heated to constant mass to drive off any moisture. When this is done, there should be no change in the crucible's mass when it is subsequently heated along with a sample. If a crucible cover is needed, it should be heated to constant mass along with the crucible.

To heat the crucible (and cover, if needed) to constant mass, insert the crucible (and cover) into a clay triangle on a support ring (see Figure 22). To avoid getting oils or residues from fingers on the crucible and cover, use crucible tongs. Heat the crucible and cover with a hot flame until they glow dull red. Cool the crucible and cover to room temperature on the clay triangle or in a desiccator. Your laboratory instructor will indicate if you are to use a desiccator.

After the crucible has been heated to constant mass and allowed to cool, put the solid to be heated into the crucible. To heat a sample so that it does not react with atmospheric oxygen, insert the covered crucible upright in the clay triangle, with the cover on straight, as shown in Figure 22(a). If you want the heated solid to react with atmospheric oxygen, tilt the crucible slightly in the clay triangle, and use crucible tongs to position the cover so it is slightly ajar, as shown in Figure 22(b).

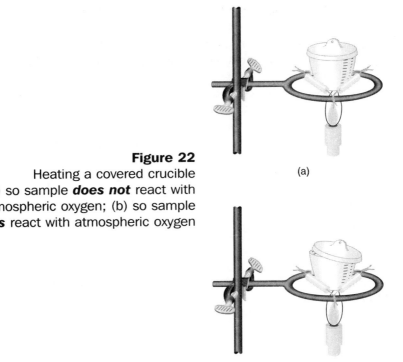

(a)

Figure 22
Heating a covered crucible
(a) so sample ***does not*** react with atmospheric oxygen; (b) so sample ***does*** react with atmospheric oxygen

(b)

Questions Regarding Heating Sources and Techniques

1. When is it advantageous to use a Bunsen burner rather than a hot plate?

2. When is it necessary to use a hot plate rather than a Bunsen burner?

3. When heating a liquid in a test tube using a burner flame, why is it necessary to: (a) move the tube in and out of the flame; (b) have the test tube no more than one-third full; (c) not heat the bottom of the test tube?

4. When is it advisable to use a hot-water bath rather than a hot plate or Bunsen burner?

5. When a burner flame goes out, why is it important to turn off the gas and wait at least 30 seconds before relighting the burner?

6. Explain why a crucible cover sometimes covers the crucible during heating and is tilted slightly at other times.

7. List the laboratory equipment shown in Figure 5(a) and (b) that you would need to convert $MgSO_4 \cdot 7 H_2O(s)$ to $MgSO_4(s)$.

8. Which pieces of laboratory equipment in Figure 5(a) and (b) would be required to determine the boiling point of a liquid? Identify any other pieces of equipment not shown in the figure that would also be needed.

VII

Separations

Decantation

A solid formed in a solution during a reaction is called a **precipitate.** The precipitate can be separated from the solution by **decantation** in which the liquid is poured off from the solid, as shown in Figure 23. To do so, first allow the solid to completely settle to the bottom of the container. The remaining liquid is called the **supernatant liquid,** or **supernate.** Then, carefully decant the supernatant liquid without disturbing or transferring any of the solid.

Figure 23
Decanting a liquid from a solid

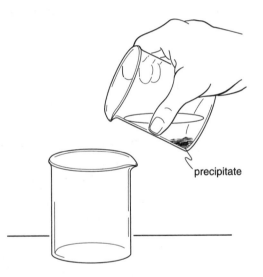

precipitate

Filtration

A precipitate can also be filtered from a solution when the solid has settled to the bottom of the reaction container. Two types of filtration are discussed here: **gravity filtration** using a conical funnel; and **suction filtration** using a partial vacuum and a Büchner funnel.

Gravity Filtration: Using a Conical Funnel

Stabilize a conical funnel by putting it into a support ring clamped to a ring stand. Select an appropriate grade filter paper with a radius that is slightly less than the length of the funnel cone [see Figure 24(a)]. Fold the paper as shown in Figure 24(b–d). Note especially that the second fold does not quite fold the paper into equal quarters. Instead, fold the paper so that the two straight edges form a 5–10 degree angle. Remove a corner of the smaller fold to provide a tighter fit of the filter paper with the funnel cone and better liquid flow during filtration. Save the

© 2002 Wadsworth Group

removed corner if its mass must be measured. Open the larger section of the folded paper to form a cone and insert it into the funnel, as shown in Figure 24(f).

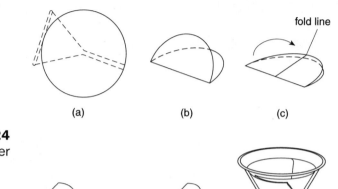

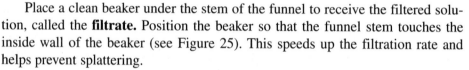

Figure 24
Folding filter paper

Place a clean beaker under the stem of the funnel to receive the filtered solution, called the **filtrate.** Position the beaker so that the funnel stem touches the inside wall of the beaker (see Figure 25). This speeds up the filtration rate and helps prevent splattering.

Press the filter paper against the funnel cone to seat the paper. Use distilled water from a wash bottle to moisten the filter paper. Add enough water to make the paper adhere tightly to the sides of the funnel and to fill the funnel stem. A tight filter paper fit and a continuous flow of filtrate speed filtration considerably.

Begin the filtration by decanting as much of the liquid as possible from the precipitate into the funnel. Use a glass rod to guide the liquid flow from the beaker to the funnel to avoid splattering, as shown in Figure 25(a). Do not allow the filter paper cone to become more than three-quarters full.

Use a glass rod fitted with a rubber policeman to scrape the precipitate from the beaker into the funnel, as shown in Figure 25(b). Use a wash bottle to rinse any precipitate remaining on the rubber, so that the rinses go into the filter paper cone.

After transferring as much of the precipitate as possible, rinse the inside of the beaker using a wash bottle, as shown in Figure 25(c). Use a glass rod to guide the rinses into the funnel.

The precipitate on the filter paper must usually be washed. Use a gentle stream of distilled water from the wash bottle to wash the precipitate down the sides of the filter paper cone to the bottom. Washing with several small portions of water is more efficient than washing with one large portion. Allow the filtration apparatus to stand until all of the liquid has passed through the filter paper into the beaker.

The next step depends on the particular procedure and the specific purpose for separating the precipitate from the filtrate. In cases where the precipitate is needed for subsequent steps, it is saved. When the filtrate is the material of interest, this liquid is saved and the precipitate discarded.

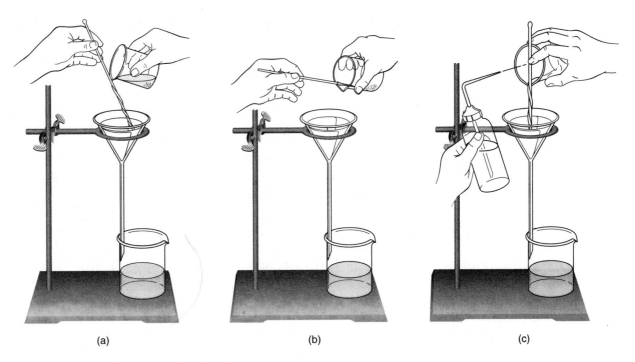

(a)   (b)   (c)

Figure 25   Proper gravity-filtration technique

Suction Filtration: Using a Büchner Funnel

Büchner funnels are made of porcelain, glass, or plastic [see Figure 26(a)]. A piece of filter paper or a glass fiber disk placed on top of the perforations in the funnel prevents the filtered precipitate from passing through the funnel. Suction from a water aspirator creates a partial vacuum that accelerates the filtration rate.

The Büchner funnel's stem is inserted into a one-hole rubber stopper that fits a *clean* side-arm filter flask used for suction filtration, as shown in Figure 26(b). Heavy-wall rubber tubing prevents tubing collapse during suction. A trap prevents accidental backflow of water from the aspirator into the filter flask. Your laboratory

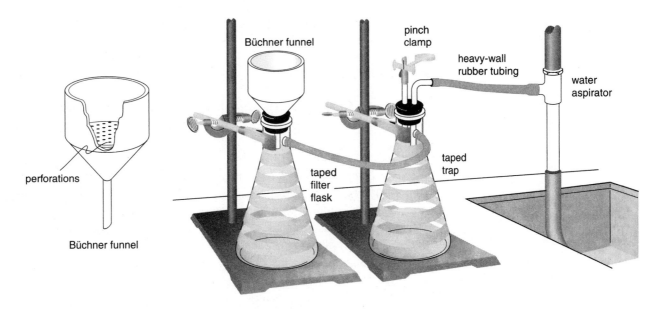

Figure 26   Suction filtration apparatus

instructor may ask you to tape the filter flask and trap to prevent glass from scattering in case the flask or trap should implode during filtration.

To prepare a Büchner funnel for use in suction, or vacuum, filtration, place a circular filter paper or glass fiber disk on top of the perforations in the funnel. The filter paper or disk should be small enough to lie flat, yet large enough to completely cover the perforations. Moisten the paper or disk with a small amount of distilled water from a wash bottle.

Create suction by turning on the water tap connected to the aspirator. Make sure that the pinch clamp is closed so that a partial vacuum can be created in the filter flask. This will cause the greater air pressure above the funnel to force the liquid in the funnel through the filter paper or disk. The pinch clamp can be used to control the rate of liquid flow through the funnel. Do **not** turn off the water tap until filtration and washing are complete, and the pinch clamp has been removed.

Once suction has begun, the vacuum filtration process is similar to gravity filtration: the bulk of the liquid is filtered first, then the bulk of the precipitate, and finally the rinse solutions. *Some liquid should be present in the Büchner funnel throughout the filtration.* After all the precipitate has been washed, continue to draw air through the precipitate to dry it somewhat, unless you are working with a precipitate that reacts with atmospheric carbon dioxide. Finally, remove the pinch clamp, and **then** turn off the tap.

Centrifuging to Separate a Precipitate from a Solution

Although a precipitate can be filtered from a solution, it is often faster and just as effective to use a centrifuge. The great advantage of centrifuging over filtration is separation speed. In most cases, centrifuge separations can be completed in about one minute, although some light, fluffy precipitates may require several minutes.

A **centrifuge** [Figure 27(a)] is a device that spins solid–liquid mixtures at high speeds in order to separate the mixture components by their mass differences into two distinct layers, as shown in Figure 27(b). Samples are held in small test tubes called **cuvets,** or similar containers. These containers are rotated at high speed that exerts as much as 300–400 times the force of gravity on the precipitate, which quickly packs the precipitate into the bottom of the sample tube. When a relatively tiny amount of precipitate must be separated, a tapered centrifuge tube is used [see Figure 27(b)]. Sample tubes should not be filled more than three-quarters full.

These precautions **must** be observed when using a centrifuge. *The centrifuge cover must be closed while the centrifuge is operating.* The centrifuge head rotates at speeds in excess of 1000 rpm. Centrifuge tubes should be examined carefully before use for possible flaws. Hands and clothes must be kept clear of the spinning parts.

The centrifuge must be kept balanced at all times. Thus, sample tubes must always be inserted in pairs, one containing the solution to be centrifuged, and the other containing an equal volume of water. The tubes must be placed directly opposite each other in the centrifuge head. Centrifuge heads usually have four or more sample compartments. When sharing a centrifuge with someone, be sure that all sample tubes are properly labeled, so that they can be identified after the centrifuge head stops spinning.

After centrifuging a sample, use a Beral or Pasteur pipet to separate the supernatant liquid from the precipitate. Squeeze the pipet bulb before inserting the pipet tip into the supernatant liquid. Remove the liquid in small portions. In some cases, centrifugation packs the precipitate so tightly into the bottom of the tube that the supernatant liquid can simply be decanted.

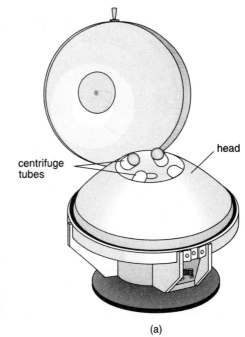

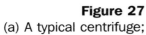

Figure 27
(a) A typical centrifuge;
(b) non-tapered and tapered
centrifuge tubes

(a)

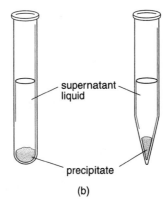

(b)

Questions Regarding Separations

1. Explain the difference between: (a) a precipitate and a supernate; (b) gravity filtration and suction filtration; (c) centrifugation and suction filtration; (d) decantation and centrifugation.

2. Explain the purpose of the trap in a suction filtration.

3. A student in a hurry decided to centrifuge a sample, so he put the sample tube into the centrifuge and turned on the machine. What mistake did the student make?

4. A mixture contains two components, A and B. A student, who needs to analyze component A, separated the components by first using a solvent that dissolves component A, but not component B. She then used suction filtration to remove B from A, and discarded the liquid that passed through the filter. What error did she make?

5. A precipitate is formed in a chemical reaction, and you want to remove the solid from the solution. List the pieces of laboratory equipment you would need and describe how they should be assembled to carry out the task. Name any other pieces of equipment not shown in Figure 5(a–b) that would also be needed.

6. In suction filtration, when is it not proper to continue to draw air through a solid in the Büchner funnel after all the liquid has been drawn from the solid?

7. One student told another to "Decant the precipitate and centrifuge the filtrate." Identify the mistakes in this advice.

VIII

Small-Scale Operations

Small-Scale Equipment

Many experiments, especially those involving solutions, can be performed using small-scale or microscale equipment. A typical small-scale experiment uses at most 3 mL of solution. Two pieces of equipment are common to most small-scale work.

The first is a **wellplate** containing small depressions called **wells** in which reactions are carried out [see Figure 28(a) on page 54]. Wellplates with 24 wells, each with a capacity of about 3.5 mL, are the most common type. The wells are identified by letters for the rows and numbers for the columns. Wellplates are also available with 96 wells, each holding about 0.5 mL, and others have fewer, larger wells.

The second important piece of small-scale equipment is a small pipet, either a plastic Beral pipet or a glass Pasteur pipet with a latex bulb [see Figures 28(b) and 28(c)]. These pipets are used to transfer liquids, dropwise if necessary, to wells of a wellplate. Beral or Pasteur pipets can also be used to obtain gas samples from a reaction mixture.

To use a Beral or Pasteur pipet to transfer liquid, first squeeze the bulb to expel air. Still squeezing the bulb, place the tip of the pipet well below the surface of the liquid and keep it there to avoid drawing air into the pipet. Slowly release pressure on the bulb to draw liquid into the pipet. Adjust the filling rate by varying the amount of pressure on the bulb. Remove the pipet tip from the liquid, releasing any remaining pressure on the bulb. Position the pipet tip over the appropriate well in the wellplate and squeeze the bulb to release the liquid into the well.

Beral or Pasteur pipets can also be used quantitatively by counting the number of drops transferred. To do so reproducibly, you must hold the pipet vertically, in order to deliver drops of uniform size and volume [see Figure 28(d)].

Small-scale procedures may also use small (12 × 75-mm) test tubes or 25-mL Erlenmeyer flasks.

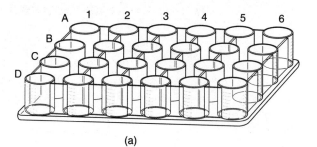

(a)

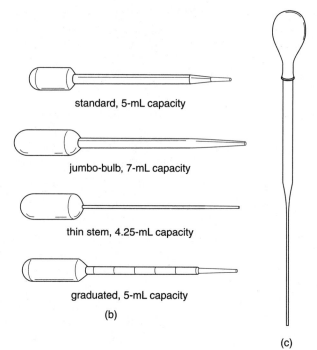

Figure 28
Two important types of small-scale equipment: (a) a 24-well wellplate; (b) four kinds of Beral pipets; (c) a Pasteur pipet with latex bulb; (d) quantitative transfer using a Beral pipet

standard, 5-mL capacity

jumbo-bulb, 7-mL capacity

thin stem, 4.25-mL capacity

graduated, 5-mL capacity

(b)

(c)

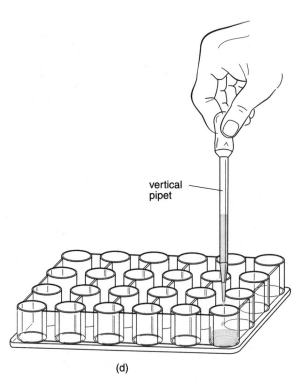

vertical pipet

(d)

Heating Small-Scale Equipment

Small-scale equipment can be heated using four different apparatus: a microburner, a hot-water bath, a sand bath, or an aluminum heating block. The microburner and hot-water bath are described in Chapter 6.

A **sand bath** is a nonflammable heat source consisting of a layer of sand in a container on top of a hot plate [see Figure 29(a)]. The temperature of a sample heated in a sand bath varies with the depth to which the sample is immersed in the sand and the hot plate temperature.

An **aluminum heating block** is a rectangular piece of aluminum metal with various-size depressions for holding glassware [see Figure 29(b)]. The block is placed on top of a hot plate, and the glassware to be heated are placed into the appropriate-size depressions. The hot plate then heats the block to the desired temperature.

In addition, a wellplate can be mildly heated by floating it in a tray of warm water.

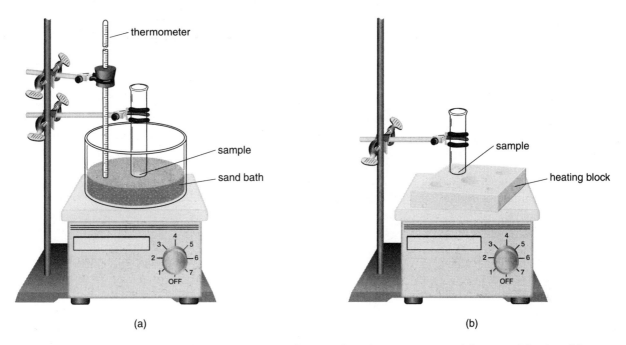

(a)   (b)

Figure 29   Two kinds of small-scale equipment heating apparatus: (a) a sand bath; (b) an aluminum heating block

Questions Regarding Small-Scale Operations

1. Why must a Pasteur or Beral pipet be held vertically when transferring drops quantitatively?

2. A sample tube deeply imbedded in a sand bath will be heated to a higher temperature than one embedded less deeply. Explain.

3. Vigorous stirring of liquids in a wellplate is to be avoided. Explain.

4. How are the positions of wells in a wellplate identified?

5. Why must the tip of a pipet be kept below the surface when drawing liquid into the pipet?

6. Why should you expel air from a pipet bulb before putting the pipet tip into a liquid?

IX

Visible Spectrophotometers

Visible absorption spectroscopy is an analytical technique commonly used to determine the concentration of a particular chemical species in a solution. The instrument used for this procedure is a **visible spectrophotometer,** which passes a beam of visible light through a sample and measures the amount of light the sample absorbs or transmits.

Visible light makes up the region of the electromagnetic spectrum that is visible to the human eye. Visible light ranges from violet to red, with wavelengths from 4×10^{-5} cm to 7×10^{-5} cm, respectively.

Substances absorb only particular wavelengths. The wavelength most strongly absorbed is referred to as the **analytical wavelength** for that substance. The color we see in a solution is the color of all wavelengths of visible light *transmitted* by the solution. In other words, we observe the complementary colors of the wavelengths *absorbed* by the solution. Thus, an aqueous bromothymol blue indicator solution appears blue because it absorbs light in the yellow region of the visible spectrum. Blue is the complementary color of yellow, the absorbed color. Table 1 lists by wavelength range the absorbed and corresponding observed (complementary) colors of solutions.

The Spectronic 20 series of instruments is the most common type of visible spectrophotometer used in general chemistry. Directions for its use follow. More detailed information regarding the nature and use of this instrument can be found in standard analytical chemistry textbooks. Check with your laboratory instructor for directions if you will be using a different model spectrophotometer.

The Spectronic 20 is used to quantitatively analyze color-absorbing substances in solutions. The principal components of a Spectronic 20 are shown in Figure 30 on page 60. You should refer to this figure as you operate the instrument.

The Spectronic 20 uses a tungsten-filament light to generate visible wavelength radiation in the 340 to 625-nm range. The wavelength control knob adjusts a diffraction grating, causing a selected wavelength of light (the incident light) to pass through a sample in solution. If the sample contains no color-absorbing substances,

Table 1   Absorbed and complementary (observed) colors of solutions

Wavelength Range Absorbed, nm	Absorbed Color	Complementary (Observed) Color
400–435	Violet	Yellow–green
435–480	Blue	Yellow
480–490	Blue–green	Orange
490–500	Green–blue	Red
500–560	Green	Purple
560–580	Yellow–green	Violet
580–595	Yellow	Blue
595–650	Orange	Blue–green
650–750	Red	Green–blue

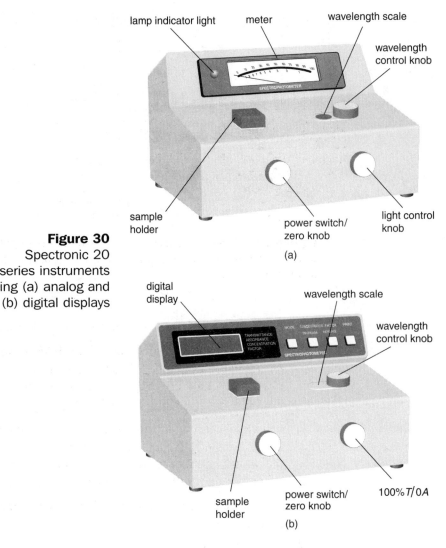

Figure 30
Spectronic 20
series instruments
showing (a) analog and
(b) digital displays

the light will pass through the sample without any of it being absorbed [Figure 31(a)]. The sample is transparent to the selected wavelength of light, and the meter reading will be 100%*T* (transmittance), which equals zero 0*A* (absorbance).

If the sample contains a color-absorbing substance, the sample will absorb light at wavelengths specific to that substance. Therefore, the amount of light transmitted will be reduced as it passes through the sample, as shown in Figure 31(b) and (c). The amount of light absorbed (or transmitted) is directly proportional to the concentration of the color-absorbing substance in the solution.

The transmitted light strikes a detector, which converts the intensity of the light to an electrical signal. In turn, this voltage is converted to absorbance or percent transmittance and displayed on an analog meter or digital display.

Refer to Figure 30 as you use your Spectronic 20, according to the instructions that follow.

1. Turn the power switch/zero knob clockwise until it clicks and the lamp indicator light comes on. (Digital models do not have such a light.) Older instruments may need to warm up for five minutes, whereas new or reconditioned instruments can be used immediately. Your laboratory instructor will advise you in this regard.

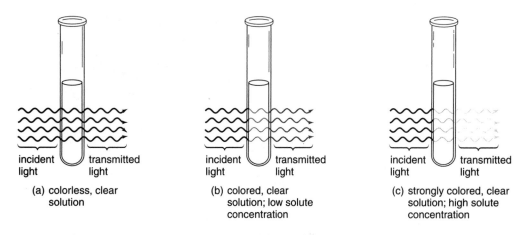

incident | transmitted
light | light

(a) colorless, clear
solution

incident | transmitted
light | light

(b) colored, clear
solution; low solute
concentration

incident | transmitted
light | light

(c) strongly colored, clear
solution; high solute
concentration

Figure 31 The absorption of visible (incident) light passed through a solution is directly proportional to the concentration of the color-absorbing substance in the solution

2. Adjust the wavelength control knob to the desired analytical wavelength.

3. Close the cover on the sample holder. *With no sample in the sample holder,* adjust the power switch/zero knob clockwise until the meter or digital display indicates a reading of 0%*T* (percent transmittance).

4. Obtain an empty test tube or cuvet, being sure to handle only the rim to avoid leaving fingerprints on the lower portion of the tube, which could affect the measurement. Half fill the tube with the **reference solution,** or **blank.** Wipe the outside of the tube with a soft tissue, such as lens paper, to remove any smudges or stray drops of solution.

5. Place the cuvet containing the reference solution into the sample holder and close the cover.

6. Adjust the light control knob until the meter or display reads 100%*T.*

7. Remove the reference solution cuvet from the sample holder. Save the reference solution in its cuvet so that you can use it to periodically recheck the light control setting.

8. Half fill a clean cuvet with a sample of the solution you wish to measure. Wipe the outside of the tube with a soft tissue to remove any smudges or stray solution. Place the cuvet into the sample holder, and close the cover.

9. Read and record the %*T* value indicated on the meter or display.

10. Repeat Steps 8 and 9 to analyze additional samples of the solution.

11. When you finish recording your measurements, turn off the instrument *if no one else is waiting to use it.* To do so, turn the power switch/zero knob counterclockwise until it clicks. The lamp indicator light will go out.

12. Clean and return all glassware as directed by your laboratory instructor. Dispose of your samples as directed.

13. If necessary, convert your %*T* values to equivalent absorbance (*A*) values, using the relationship

$$A = 2.00 - \log \%T \text{ or } A = -\log (\%T/100)$$

Questions Regarding the Spectronic 20 Series

1. What is the purpose of the blank?

2. Why is it important to set the instrument to 100% transmittance using the blank?

3. Why is it important to set the instrument to 0% transmittance?

4. Why must the cover of the sample holder be closed while evaluating a sample?

5. The maximum absorbance of a particular solution occurs at 540 nm. What color is the solution?

6. Suppose the wavelength control was incorrectly set to a wavelength at which the sample being analyzed had very low absorbance. How would this error affect the analysis of solute concentration in this solution?

X

Using a pH Meter

A **pH meter** measures the acidity (H_3O^+ concentration) of an aqueous solution and converts it to a pH value. (Actually, a pH meter measures the *activity* of H^+, although H^+ activity is closely related to H_3O^+ concentration.) While there are many different pH meter designs, all models basically consist of a voltmeter with attached electrodes (Figure 32). Every pH meter has two electrodes, but in many cases the two are in one unit, called a **combination electrode,** as shown in Figure 32. The two electrode types, whether separate or combined, are: a **reference electrode,** which has a constant voltage; and an **indicator electrode,** generally a glass electrode, which develops a voltage that is dependent on the H_3O^+ concentration (H^+ activity) of the sample being analyzed. The pH meter measures the difference between the voltages of the indicator and reference electrodes, converts this difference to a pH value, and displays this value.

The following steps outline the use of an AC-powered benchtop pH meter, with the features illustrated in Figure 32.

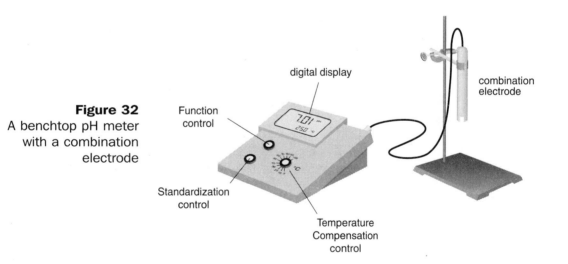

Figure 32
A benchtop pH meter with a combination electrode

1. **Getting Started** If the combination electrode has a removable protective cap, remove it. Carefully submerge the tip of the electrode in a beaker containing distilled or deionized water. Turn the function control to "standby", and plug the power cord into a 110-V AC circuit. Turn the function control to "pH"; a pH value should appear on the display. If no pH value appears, turn the function control back to "standby", and consult your laboratory instructor.

2. **Standardizing the Meter** To measure pH values accurately, a pH meter must be calibrated using reference standards, as described in the following steps. Choose a reference standard whose pH is close to that of the solution

you wish to analyze. If the pH meter is to be used for an experiment with a wide-ranging pH, such as an acid–base titration, it is helpful to use two reference standards, one with a pH near the midpoint of the experimental range, and another with a higher pH, near to one that is likely to occur at the end of the experiment.

Calibration Procedure

a. Rinse the tip of the electrode with distilled or deionized water, collecting the rinses in a labeled beaker.

b. If the electrode is fillable, twist the top of the electrode to uncover the fill hole.

c. Carefully submerge the tip of the electrode in the reference solution. Make sure that the glass bulb is covered by the solution. ***Caution:*** The glass bulb is fragile.

d. If your pH meter does not have an automatic temperature control, measure the temperature of the reference solution. Adjust the temperature compensation control to match the solution temperature.

e. Turn the function control to "pH" and check to see that a pH value appears on the display.

f. Adjust the standardization control until the display indicates the exact pH of the reference solution, being sure to wait several seconds until the reading stabilizes. You should ***not*** adjust the standardization control during subsequent pH measurements. If you are using two reference standards, repeat steps **2** (c)–(f) using your second reference solution.

g. Turn the function control to "standby". Carefully remove the electrode from the reference solution and rinse it with distilled water. Collect the rinses in a labeled beaker.

3. Measuring the pH of a Solution

a. Carefully submerge the tip of the electrode in the solution you wish to analyze. Gently swirl the solution so that it makes good contact with the electrode tip. Do not bump the electrode against the sides or bottom of the beaker.

b. Turn the function control to "pH" and read the displayed pH. Record the pH value.

c. Turn the function control to "standby". For subsequent pH measurements, repeat steps **3** (a)–(b), turning the function control to "standby" between measurements, if different samples are analyzed. Be sure to rinse the electrode with distilled water between samples.

d. After your final pH reading, turn the function control to "standby". Carefully remove the electrode from the solution and rinse it with distilled water. Collect the rinses in the discard beaker.

e. Submerge the electrode tip in a reference solution supplied by your laboratory instructor. If the electrode is fillable, twist the cap to cover the fill hole.

© 2002 Wadsworth Group

Questions Regarding Use of a pH Meter

1. Why must you calibrate a pH meter with a reference standard?

2. Why should two reference standards be used to calibrate a pH meter being used to monitor an acid–base titration?

3. A pH meter has been called a "special voltmeter". What is the relationship between the voltage measured and the pH value displayed?

4. A student needs to accurately determine the pH of a series of samples whose pH ranges from 8 to 11. The student calibrates the pH meter using a pH 4 reference standard. Comment on the student's choice of reference standard.

5. Explain why a pH meter has both a reference and an indicator electrode.

XI

Keeping a Laboratory Notebook

Keeping a laboratory notebook is not required by every laboratory instructor. However, the habit of recording laboratory activities is a useful one. Writing down information pertinent to an experimental procedure can be an excellent reminder of previous work as the experiment proceeds. Conscientiously recording entries as laboratory work proceeds makes the laboratory notebook a permanent record of experimental observations and data, and allows a review of experimental progress.

Because the laboratory notebook is a permanent record, it should have a permanent binding, and entries should be made in ink. When a data entry error occurs, draw a single line through the incorrect data so that the incorrect information can be read later, if necessary. Figure 33 on pages 70 and 71 shows two representative laboratory notebook pages.

On the first notebook page, record your name, locker number, course name and number, and any other important facts concerning your laboratory section. Leave the next several pages blank; later, they can become the table of contents for the notebook. Use left-hand pages for raw experimental data and calculations. Use right-hand pages for the write-ups of the experiments. Each write-up should include the experiment title, the date of the experiment, a summary of the experiment, its purpose and procedure, a record of observations including tables and graphs, an analysis of the observations and data, and conclusions (see Figure 33).

There are notebooks that contain a carbonless copy of each page; therefore, there are no left-hand or right-hand pages. In such notebooks, the original page stays bound in the notebook, while the copy can be torn out and handed in. In this style of notebook, divide the page vertically in two; the left portion is used for data and calculations, and the right portion for the experimental write-up.

Figures are a graphical way to describe a procedure or an experimental setup. They can be used to emphasize a laboratory technique, indicate a specific location on a piece of apparatus, or otherwise help to present an idea more clearly. Figures sketched in a laboratory notebook can be useful references for later experiments. Many of the figures in this handbook illustrate these points.

January 12, 2001

HCl used:

$(2.500 \times 10^{-2}\,L)\,(5.880 \times 10^{-1}\,mol/L) = 1.470 \times 10^{-2}\,mol\,HCl\,added$

NaOH used, back titration: Trial 1, 5.1 mL

$mol\,NaOH = (5.10 \times 10^{-3}\,L)\,(9.83 \times 10^{-2}\,mol\,NaOH\,/L)$

$= 5.01 \times 10^{-4}\,mol\,NaOH$

$= 5.01 \times 10^{-4}\,mol\,HCl\,not\,neutralized\,by\,Tums\,tablet$

mol HCl neutralized by Tums tablet =

$(1.470 \times 10^{-2}\,mol\,HCl\,added) - (5.01 \times 10^{-4}\,mol\,HCl\,not\,neutralized\,by\,tablet)$

$= 1.42 \times 10^{-2}\,mol\,HCl\,neutralized/tablet$

$= (1.42 \times 10^{-2}\,mol\,HCl/tablet)\left(\dfrac{1\,tablet}{1.250\,g}\right) = 1.14 \times 10^{-2}\,mol\,HCl/g$

Figure 33   Entering data in a laboratory notebook: raw data and calculations go on left-hand pages

January 12, 2001

COMMERCIAL ANTACID ANALYSIS

SUMMARY

Tums antacid tablets were analyzed for relative effectiveness in neutralizing a standardized HCl solution. The Tums data were compared with those of other commercial antacids.

PURPOSE and PROCEDURE

Tums antacid tablets were analyzed for relative effectiveness by reaction with an excess of standardized 0.5880M HCl. The excess of HCl was then back titrated with standardized 0.0983M NaOH to determine the amount of HCl neutralized by each tablet. The antacid effectiveness was expressed as mol HCl neutralized per gram of tablet.

OBSERVATIONS

Gas evolved as the tablet reacted with HCl. After gas evolution stopped, a murky white solution remained.

Addition of bromphenol blue indicator turned the solution yellow. The end point in back titration was reached when the yellow solution turned green.   1.184

	Trial 1 — 1.250 g tablet	Trial 2 — ~~1.187~~ g tablet
Volume of 0.0983M NaOH:		
	Final = 5.39 mL	Final = 10.57 mL
	Initial = 0.29 mL	Initial = 5.39 mL
	Used = 5.10 mL	Used = 5.18 mL

ANALYSIS OF OBSERVATIONS AND DATA

The evolved gas was CO_2 generated by the reaction with the antacid, likely HCO_3^- or CO_3^{2-} in the tablet:

$$H^+(aq) + HCO_3^-(aq) \rightarrow H_2O(\ell) + CO_2(g)$$

Figure 33 (cont.)   Use right-hand pages for formal write-up of experiments

XII

Mathematical Operations

Exponential Notation

Studying chemistry involves using very large and very small numbers. For example, 1 mL of water contains more than 30,000,000,000,000,000,000,000 molecules. Each water molecule has an approximate mass of 0.000 000 000 000 000 000 000 03 g, far too small for any balance to measure the mass. Representing very large and very small numbers in this way is awkward and time consuming. Instead, we use **exponential notation,** also called **scientific notation,** to express such numbers.

Expressing Numbers Using Exponential Notation

Exponential notation expresses numbers as the product of two factors: (1) the **digit term,** a number between 1 and 10; and (2) the **exponential term,** which has the form 10^x, that is, 10 raised to a whole-number power called an **exponent.** By convention, the digit term has just one number to the left of the decimal point. For example, we can express 123 as 1.23×10^2, read as "one point two three times ten to the second." The 1.23 is the digit term and 10^2 is the exponential term.

As shown in the following examples, numbers greater than 10 have positive exponents; those less than one have negative exponents.

$$273.15 = 2.7315 \times 10^2$$

$$0.0415 = 4.15 \times 10^{-2}$$

$$0.001 = 1 \times 10^{-3}$$

The exponent is equal to the number of places we must move the decimal point to convert the number into the digit term, one with just a single number to the left of the decimal point. If the decimal point must be moved to the *left,* the exponent is positive; for example, $273.15 = 2.7315 \times 10^2$. The decimal point in 273.15 must be moved two places to the left to produce the digit term 2.7315, so the exponent is 2. To convert 0.0415 to the digit term 4.15, the decimal point must be moved two places to the *right.* Consequently, the exponent is –2, and the exponential notation for 0.0415 is 4.15×10^{-2}.

Exponential Notation Using a Calculator

Your calculator may be slightly different from the one used for the following examples. If so, use the instructions provided with your calculator when performing these tasks. Your calculator must have an exponent key, usually labeled EXP (or EE or EEX).

1. **Entering Exponential Numbers** To enter 1.23×10^2, use the keys in the following order: ⌨1⌨ ⌨.⌨ ⌨2⌨⌨3⌨ ⌨EXP⌨ ⌨2⌨ . To enter 4.15×10^{-2}, the order is ⌨4⌨ ⌨.⌨ ⌨1⌨⌨5⌨ ⌨EXP⌨ ⌨±⌨ ⌨2⌨ . The ⌨±⌨ key may be labeled CHS for "change sign". Some calculators can be programmed so that numbers are automatically expressed in exponential form. Check the instructions with your calculator to see whether this option is available.

2. **Adding, Subtracting, Multiplying, and Dividing Exponential Expressions** Use the ⌨+⌨, ⌨−⌨, ⌨×⌨, or ⌨÷⌨ keys, respectively, to carry out these operations. To multiply 3.2×10^{-3} by 5×10^{-4}, use these keystrokes.

   ⌨3⌨ ⌨.⌨ ⌨2⌨ ⌨EXP⌨ ⌨±⌨ ⌨3⌨ ⌨×⌨ ⌨5⌨ ⌨EXP⌨ ⌨±⌨ ⌨4⌨ $= 1.6 \times 10^{-6}$

3. **Determining Square Roots and Cube Roots of Exponential Numbers** For square roots, remember that $\sqrt{A} = A^{1/2}$, and use the ⌨√x⌨ or ⌨yˣ⌨ key. For calculators with a ⌨yˣ⌨ key, use the following sequence to obtain the square root of 2.7×10^8:

   ⌨3⌨ ⌨.⌨ ⌨2⌨ ⌨EXP⌨ ⌨±⌨ ⌨3⌨ ⌨×⌨ ⌨5⌨ ⌨EXP⌨ ⌨±⌨ ⌨4⌨ $= 1.6 \times 10^4$

The ⌨.⌨ and ⌨5⌨ keys are used because $1/2 = 0.5$.

Exponential Notation Without Using a Calculator

1. **Multiplying Exponential Numbers** First multiply the digit terms, then **add** the exponents to determine the exponential term of the answer. In general,

$$(A \times 10^x)(Z \times 10^y) = (A \times Z) \times 10^{x+y}$$

For example:

$$(2 \times 10^3)(3 \times 10^2) = (2 \times 3) \times 10^{3+2} = 6 \times 10^5$$
$$(3 \times 10^{-2})(2 \times 10^{-4}) = (3 \times 2) \times 10^{(-2)+(-4)} = 6 \times 10^{-6}$$

2. **Dividing Exponential Numbers** First divide the digit terms, then **subtract** the exponents to determine the exponential term of the answer. In general,

$$\frac{A \times 10^x}{Z \times 10^4} = \frac{A}{Z} \times 10^{x-y}$$

For example:

$$\frac{8 \times 10^3}{4 \times 10^2} = \frac{8}{4} \times 10^{3-2} = 2 \times 10^1$$

$$\frac{8 \times 10^3}{4 \times 10^{-2}} = \frac{8}{4} \times 10^{(3)-(-2)} = 2 \times 10^5$$

3. **Adding and Subtracting** Numbers must have the same exponents in order to add them or subtract them without a using a calculator. To get the same exponents, move the decimal point(s) and adjust the exponent accordingly, so that the exponents are alike. To add 1.23×10^{-2} and 2.34×10^{-1}, move the decimal point to convert 1.23×10^{-2} to 0.123×10^{-1}. Then carry out the addition: $0.123 \times 10^{-1} + 2.34 \times 10^{-1} = 2.46 \times 10^{-1}$. We also could have changed 2.34×10^{-1} to 23.4×10^{-2}; after the addition, $23.4 \times 10^{-2} + 1.23 \times 10^{-2}$, the answer is 24.6×10^{-2}, or 2.46×10^{-1}.

   The rules for subtraction are the same as those for addition, except that the digit terms are subtracted, not added.

© 2002 Wadsworth Group

Significant Figures

Every measurement involves some degree of error. **Significant figures** are a way of expressing the degree of confidence in a measurement; that is, the estimated degree of error. The more significant figures included in a measurement, the higher the level of confidence in that measurement.

A reasonably reliable measurement contains at least one figure known with certainty, plus an estimated figure to the right of the last known figure. Only one estimated figure is included in the reported number of significant figures. The following general rules apply to significant figures.

1. All nonzero numbers are considered significant figures. Thus, 1.234 has four significant figures.

2. The decimal point position has no effect on the number of significant figures. If a zero is surrounded by nonzero numerals, it is a significant figure; 1.03 has three significant figures.

3. Zeros to the left of the first nonzero numeral are not significant. For example, 0.0802 has three significant figures; 8, the third 0, and 2. The first two zeros are not significant.

4. **Trailing zeros** are those to the right of the last nonzero numeral. All trailing zeros to the right of a decimal point are significant. Thus 25.00 has four significant figures; 0.00450 has three significant figures (4, 5, and the final 0).

5. If a number is expressed in exponential notation, the number of significant figures should be equal to the number in the digit term; 0.04504 would be expressed as 4.504×10^{-2}, with four significant figures in the digit term.

Significant Figures in Addition and Subtraction

The number of decimal places in the answer should equal the number of decimal places in the factor with the *fewest* places to the right of the decimal point. For example, 0.13 + 1.534 + 6.1 = 7.764, which should be reported as 7.8 (one decimal place), because 6.1 has only one decimal place. Although 1.534 has four significant figures, three to the right of the decimal point, it is the factor with the fewest number of decimal places that dictates the number of places in the answer.

Significant Figures in Multiplication and Division

These rules are different from those for addition and subtraction. In multiplication and division, the number of significant figures in the answer should be the same as the number of significant figures in the quantity with the *fewest* significant figures. Consider 0.001319 divided by 0.0206. The first number has four significant figures, while the second has only three. Therefore, the answer is limited to three significant figures.

$$\frac{0.001319}{0.0206} = 0.0640 = 6.40 \times 10^{-2}$$

Rounding Off Numbers

Rounding off, the dropping of certain digits from a calculated result, is necessary because a calculated result can only be as reliable as the *least* precisely known factor in the calculation. The **dropped digit** is the digit to the right of the **retained digit,** the last one to be kept. For example, if 5.432 is rounded off to 5.43, then 2 is the dropped digit and 3 is the retained digit. The following general rules apply when rounding off numbers:

1. If the dropped digit is less than 5, do not change the retained digit.

2. If the dropped digit is greater than 5, increase the retained digit by 1.

3. If the dropped digit is 5 followed by nonzero digits, increase the retained digit by 1.

4. If the dropped digit is 5 followed by zeros or no digits:

    a. the retained digit remains unchanged if it is an even number;

    b. the retained digit is increased by 1 if it is an odd number.

For example,

Original Number	Number Rounded to Three Significant Figures
15.798 g	15.8 g
13.247 g	13.2 g
21.75 g	21.8 g
22.252 g	22.3 g
16.4500 g	16.4 g

Rounding Off Calculated Results

When experimental values are added or subtracted, the result can have no more decimal places than the least precisely known number being added or subtracted. When multiplying or dividing experimental results, the result can have no more significant figures than the factor with the least number of significant figures. For example,

Mass of solid before heating	24.662 g	24.7 g
Mass of solid after heating	19.3 g	19.3 g
Mass lost		5.4 g

$$\text{Mass of a methane molecule} = \frac{16.042\,\text{g}}{6.022 \times 10^{23}\,\text{molecules}}$$

← 5 significant figures
← 4 significant figures

$$= 2.664 \times 10^{-23}\,\text{g / molecule}$$ ← 4 significant figures

Dealing with Measurement Errors

Although error is unwanted, it nonetheless affects the accuracy of **all** measurements, to at least some degree. One common type of measurement error is **systematic error,** such as that caused by an incorrectly calibrated measuring device. In such cases, the error will cause consistent incorrectly high or low values, depending on the direction of the calibration error.

Random error, another common type, is caused by the fact that any measurement falls somewhere within a range of probable measurements, so it is really only an *estimate* of the actual value. Random error can be reduced by increasing the number of times each measurement is made and then calculating the **average** or **mean** of the measurement data.

Standard Deviation

The **precision** of a set of measurements indicates how well the measurements agree with each other. **Accuracy** refers to how well a particular measurement agrees with the actual value. **Standard deviation,** *s,* is a statistical measure of the precision of a set of measurements. The value of *s* is calculated using Equation 1.

$$s = \frac{\sqrt{\sum(x_i - \bar{x})^2}}{n-1} \tag{Eq. 1}$$

Several values must be determined in order to calculate a standard deviation. First find the average, $\bar{x}$, of the measured values. Then calculate the deviation, $x_i - \bar{x}$, of each of the measured values, x_i, from the average, $\bar{x}$. Square each of the deviation values, $(x_i - \bar{x})^2$, and add the results. Divide the sum of the squared deviation values by the number of measured values, *n,* minus one; $n - 1$. Finally, take the square root of the result.

The standard deviation is preceded by a ± (plus or minus) sign and is usually written after the average of the measured values to indicate the probable deviation from the true value; for example, 0.52 ± 0.003 g.

We can calculate the standard deviation of the following mass measurements: 0.61 g, 0.58 g, 0.57 g, 0.63 g, 0.62 g, and 0.59 g. The average, $\bar{x}$, is 0.60 g. The deviation of each measurement from the average, and the square of each deviation, are listed in the following table.

Measurement, g	$x_i - \bar{x}$	$(x_i - \bar{x})^2$
0.61	0.01	0.0001
0.58	−0.02	0.0004
0.57	−0.03	0.0009
0.63	0.03	0.0009
0.62	0.02	0.0004
0.59	−0.01	0.0001

The sum of the squared deviations is 0.0028. Divide this sum by $n - 1 = 6 - 1 = 5$. The square root of the result, 0.00056, is ± 0.024 g, which is the standard deviation for this set of data. This deviation can be expressed with the average as 0.60 ± 0.024 g.

XIII
Representing Data and Results Using Graphs

Graphs

Experimental data are often organized and recorded in a table that pairs the experimental conditions, the **independent variable,** with the corresponding observations and measured data, the **dependent variable** (see Figure 34). Frequently a graph helps to establish fundamental relationships between the data. Graphs are also used to predict or estimate data that are difficult to determine experimentally.

A **graph** is a pictorial diagram of data. The simplest and most widely used type for plotting scientific data is the **line graph,** an example of which is shown in Figure 34. Line graphs contain a vertical line, called the **y-axis,** or **ordinate,**

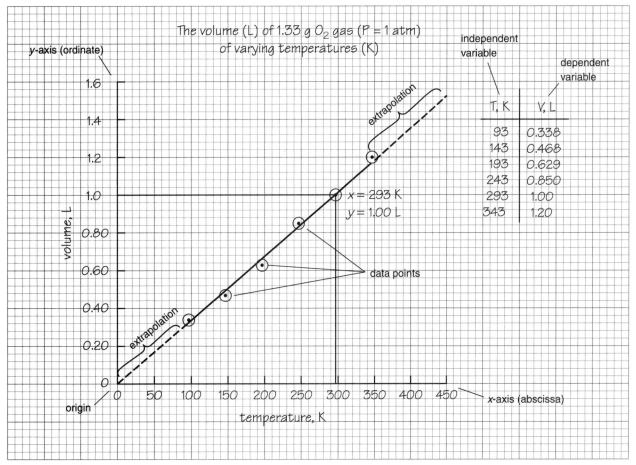

Figure 34 Data table and graph showing the volume (in liters) of 1.33 g O_2 gas (P = 1 atm) at varying temperatures (K)

and a horizontal line, called the ***x*-axis,** or **abscissa.** Each axis is divided into a numerical scale of units. Usually, the scale values increase reading upward on the *y*-axis and to the right on the *x*-axis. Numbers on the *y*-axis are usually values of the dependent variable (volume, in the case of Figure 34). Numbers on the *x*-axis normally represent values of the independent variable (temperature, in Figure 34). If the axes intersect at the point where the values represented on both axes equal 0, this point is called the **origin.** If the extrapolated line drawn through the data points does not pass through the origin, but instead crosses the *y*-axis at some other point, the intersection point is known as the ***y*-intercept.**

Each **data point** plotted on a graph represents the intersection of an *x*-axis value and the corresponding *y*-axis value from the data table for that graph. Data points are used to draw a smooth curve or straight line that best represents all of the data points. Note that the data points are ***not*** connected by a jagged line drawn directly from point to point. Rather, the **best straight line** is drawn, one that has approximately equal numbers of data points on either side of the line, with some data points on the line, as shown in Figure 34.

When preparing a graph, leave plenty of space for plotting data. Draw the axes using a straightedge. Allow space for labels along the left-hand side of the *y*-axis and under the *x*-axis.

Construct all graphs so that the lines or curves representing the data points are as long as possible. To do so, select scales for the axes that will cause the graph to extend across a full page of graph paper. This will make the graph easier to interpret.

Based on the range of the data, choose appropriate values to begin and end each axis scale, and decide on the scale-unit size for each axis. Then mark the axes with the scale units. Note that the units on the two axes do not have to be (and often are not) the same size. Generally, you should not number individual scale units, due to space considerations. Rather, select a convenient multiple of the units and mark these multiples at the appropriate intervals. For example, Figure 34 shows *x*-axis unit labels that are multiples of 50.

Label each axis with the property being plotted and the corresponding unit, as shown in Figure 34. Then plot the data points, and draw the best straight line or smooth curve through those points. Title the graph appropriately for easy identification.

Equation and Slope of a Straight Line

The general equation for a straight line is

$$y = mx + b$$

where *y* is a value on the *y*-axis, *x* is the corresponding value on the *x*-axis, *m* is the slope of the line, and *b* is the point where the line intersects the *y*-axis. For example, a plot of Fahrenheit temperature (*y*-axis) versus Celsius temperature (*x*-axis) yields a straight line for which the slope, *m*, is 1.8 °F/°C, and the *y*-axis intercept, *b*, is 32 °F.

Frequently, the slope of a straight-line graph represents an important relationship between the experimental data. For example, the slope of the line for a plot of mass versus volume measurements of several samples of a substance, g/mL, is equal to the density of that substance. The slope of a straight line can be determined by dividing the change in the *y*-axis values of two points on the line by the change in the x-axis values of those two points. For example, the slope of the line in a plot of temperature versus volume (Figure 34) can be calculated using two points on the line: $T_1 = 93$ K, $V_1 = 0.338$ L and $T_2 = 243$ K and $V_2 = 0.850$ L. The change in temperature (x-axis) is 243 K – 93 K = 150 K. The change in volume (y-axis) is 0.850 L – 0.338 L = 0.512 L. Therefore, the slope of the line can be calculated:

$$\text{slope} = \frac{0.850\,\text{L} - 0.338\,\text{L}}{243\,\text{K} - 93\,\text{K}} = \frac{0.512\,\text{L}}{150\,\text{K}} = 3.41 \times 10^{-3}\ \text{L/K}$$

Notice that the units of the slope are the units of the y-axis divided by those of the x-axis. In this case, the units of the slope are L/K, which indicates the volume change of oxygen gas with a change in temperature. If the slope is positive ($m > 0$), as in this case, the line rises from left to right. If the slope is negative ($m < 0$), the line falls from left to right, as in a plot of mass versus time for an evaporating liquid.

Graphing Programs

Computerized graphing software automatically makes graphs that would be time consuming to create manually. Graphing programs can also investigate relationships between the independent and dependent variables, beyond just the plot of these variables. A graphing calculator or computer graphing program can plot a **linear regression line,** which is the best straight line passing through, or as close as possible to, all the data points on a graph. The calculator or computer program may also be able to determine the correlation between the data points and the linear regression line. The **correlation** is a value between 0 and 1 that indicates how well the linear regression fits the data points; that is, how close the data points are to forming a straight line. A high correlation value (one close to 1) indicates a close fit between the data and the linear regression line; the plot has little **scatter** to the data. The lower the correlation value, the more scattered the data, with fewer points on or near the straight line.

Graphing Exercise

The following graphing exercise uses both manual graphing and Microsoft® Excel 2000 to plot mass versus time data for a sample of acetone, $(CH_3)_2CO$, a volatile liquid, in an open container. The data are given in the following table.

Time, min	Mass Acetone, g	Time, min	Mass Acetone, g
0.0	8.860	6.0	8.765
1.0	8.844	7.0	8.748
2.0	8.828	8.0	8.734
3.0	8.814	9.0	8.719
4.0	8.796	10.0	8.703
5.0	8.782		

Manual Graphing

1. Using these time and mass data, prepare an accurate *hand-drawn* graph of mass (y-axis) versus time (x-axis) using the graph paper provided at the end of this book.

2. Draw the best straight line through the data points, and calculate the slope of the line.

Microsoft® Excel 2000 Graphing*

Note: If Excel 2000 is not available, Excel 97 can be used. The only difference is in the **Creating the Graph** section. If you are using Excel 97, in Step 5, under "**Chart sub-type**," choose the option with unconnected points.

Three typestyles are used in the following directions: (1) **boldface** is used for system menus, categories within menus, dialog boxes, and cell references; (2) *italics* is used for menu choices and keyboard entries for menu choices; (3) UPPERCASE indicates individual keystrokes, including OK and NEXT.

Data Entry Using the Time and Mass Data for Acetone

1. Open a new file in Excel and enter a descriptive title for your graph in cell **A1.** Press ENTER on your keyboard when you are finished.

2. Enter a label and unit (*time, minutes*) in cell **A3.** Do the same for (*mass, grams*) in cell **B3.** Don't worry if the labels and units extend beyond the cell boundaries; you can correct that later. Excel assumes that the values in the left-hand data column are for the independent variable and are to be plotted on the *x*-axis.

3. Enter the time data in cells **A4–A14** and the corresponding mass data in cells **B4–B14.** To display all values to three decimal points, click in cell **A4** and drag to cell **B14** to highlight all the time and mass entries. Now select **Format** and **Cells.** A **Number** tab will appear, and with it, a **Category.** Choose *Scientific* from the **Category** menu, then *3 Decimal Places,* and click OK.

Creating the Graph

4. Click and drag to highlight cells **A4** to **B14.**

5. Click **ChartWizard** on the toolbar, or select *Chart* from the **Insert** drop-down menu. From the **Chart Type** menu, choose **Chart type:** *XY (Scatter)*; from **Chart sub-type,** select *Smooth Lines without Markers.* Click NEXT.

6. The **Chart Source Data** menu shows a preview of the plot of mass versus time. Check to make sure the menu entries read **Data range:** = *Sheet1!$A$4:$B$14* and **Series in:** *Columns.* Make any necessary corrections, then click NEXT.

7. Click the **Titles** tab and enter a descriptive **Chart title.** Enter a label and units for the *x*-axis in the **Value (X) Axis** box, and a label and units for the *y*-axis in the **Value (Y) Axis** box.

   Click the **Gridlines** tab and deselect any gridlines that have been selected by ChartWizard. Click NEXT.

8. Choose **As object in:** *Sheet 1* from the **Chart Location** menu. Click FINISH to display the graph.

---

*This exercise is adapted from one found in MISC 877, *Introduction to Computer-Based Graphical Analysis,* by M.L. Gillette, H.A. Neidig, and J.R. Crook, © 2000 by Chemical Education Resources. A more complete treatment of graphing using Microsoft® Excel 97 is contained in this title.

9. The plot should appear roughly linear. To confirm its linearity, click on any data point, using the right mouse button. This should highlight all data points, and a shortcut menu should appear. From this menu choose **Add Trendline.** From the **Add Trendline** menu, choose the **Type** tab, and select **Trend/Regression type:** *Linear.* From the same menu select the **Options** tab and click *Display equation on chart* and *Display R-squared value on chart.* Select *Automatic* from the **Trendline Name** submenu. Click OK to finish.

The line now passing through the data points is the linear regression line. The equation for the linear regression line appears inside the chart area. The equation is in the form $y = mx + b$, where m is the slope of the line. The units of the slope are the y-axis unit (grams), divided by the x-axis unit (minutes). The value of R^2 is the correlation value.

Appendix A: Operations with Glass

Cleaning Glassware

Glassware used in experiments must be clean. To clean beakers, test tubes, flasks, and other glassware, use hot water and soap or detergent. Use the appropriate brushes to clean inside the glassware. Make sure that the wash water and rinses flow through the tips of pipets and burets. Rinse the glassware with tap water two or three times, then once or twice with small amounts of distilled water. Glassware is clean after the final rinse if no water droplets cling to the washed part of the glassware. Allow the clean glassware to air dry before using it.

Working with Glass Tubing

Cutting Glass Tubing

Place the glass tubing on the benchtop. At the point where the tubing is to be cut, mark the tubing with a light scratch made with a triangular file or a glass scorer [see Figure 35(a) on the next page].

Hold the tubing so that the scratch is facing away from you [see Figure 35(b)]. Apply light forward pressure with your thumbs close to each other on the side of the tubing directly opposite the scratch. Simultaneously pull the ends of the glass tubing toward your body with your fingers, while quickly pushing your thumbs forward to snap the tubing [see Figure 35(c)].

Figure 35(d) shows the ends of two pieces of glass tubing, one cut correctly and the other cut incorrectly. When the cutting is done correctly, the cut end should appear even, not jagged.

Firepolishing Glass Tubing

Firepolishing makes the cut ends of glass tubing smooth and safe to handle. While rotating the tubing, heat the cut end in the hottest part of a nonluminous flame until the end appears smooth [see Figure 36(a) on the next page]. Do not heat the tubing so long that the opening at the cut end becomes constricted. Figure 36(b) shows the ends of correctly and incorrectly firepolished tubing. The heated part of the glass is hot enough to burn your fingers or paper. Therefore, place the hot glass on a piece of ceramic-centered wire gauze or other insulated surface to cool.

Bending Glass Tubing

A good tubing bend is smooth, with the tubing diameter remaining constant throughout the length of the bend, as shown in Figure 37 on page 91.

To bend glass tubing, place a wing top, or flame spreader, on top of a gas burner, shown in Figure 38(a) page 92. Ignite the gas, and adjust the flame so that an intense blue flame spreads along the wing top.

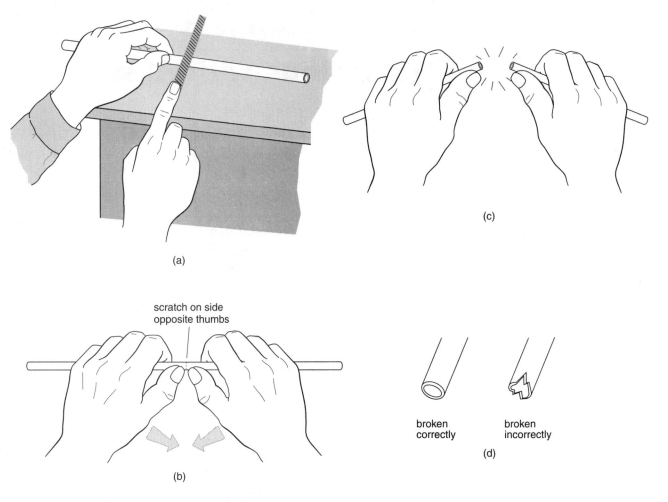

Figure 35 Cutting glass tubing: (a) scratching a piece of glass tubing; (b) correct hand positioning on the tubing; (c) snapping the tubing; (d) correctly and incorrectly cut tubing

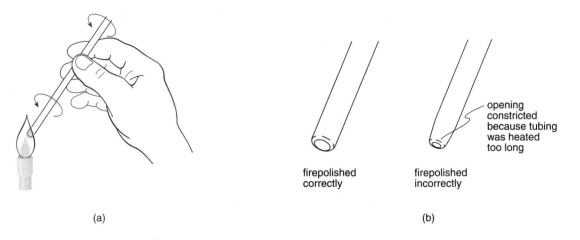

Figure 36 (a) Firepolishing glass tubing; (b) correctly and incorrectly firepolished tubing

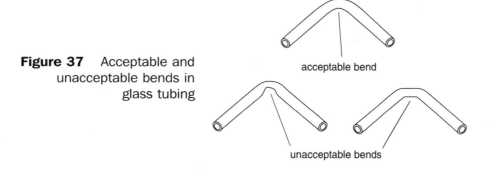

Figure 37 Acceptable and unacceptable bends in glass tubing

Obtain a piece of glass tubing that has been cut to the correct length and fire-polished. Hold the tubing on either side of where the bend is to be made. Place the part of the tubing you wish to bend just above, and parallel to, the top of the bright blue portion of the flame. Rotate the tubing slowly and continuously, as shown in Figure 38(b). Handle the hot tubing with care to avoid burning your fingers or hands.

When the glass begins to sag, remove it from the flame. Bend the tubing to the desired angle by quickly lifting the ends of the tubing toward each other, as shown in Figure 38(c). Place the bent tubing on a ceramic-centered wire gauze or other insulated surface to cool.

Turn off the gas flow to your burner. Allow the wing top and burner barrel to cool before removing the wing top from the barrel.

Preparing Glass Droppers and Pipets

Begin with a length of tubing with firepolished ends. Adjust a burner, without a wing top, to produce a hot, nonluminous flame.

Hold the tubing horizontally between your fingers and thumbs, as shown in Figure 39(a) on the next page. Keep the center of the tubing just above the bright blue inner cone of the flame. Continuously rotate the tubing in one direction in the flame.

When the tubing begins to sag, slowly pull the ends of the tubing away from each other, keeping your hands parallel to each other [see Figure 39(b)]. Quickly remove the tubing from the flame. Continue to pull the tubing, keeping your hands parallel [see Figure 39(c)]. Hold the tubing in this position until it cools and hardens.

Cut the tips to the desired size using a triangular file or a glass scorer. Firepolish the ends of the tips, heating them only ***briefly*** to avoid constricting the narrow openings. Place the firepolished droppers on a ceramic-centered wire gauze or other insulated surface to cool.

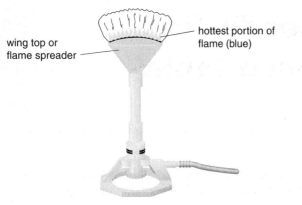

wing top or flame spreader

hottest portion of flame (blue)

(a)

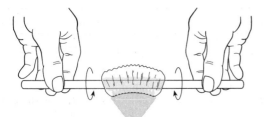

(b)

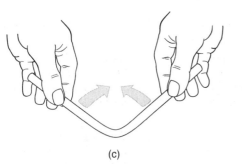

(c)

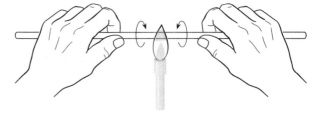

(a)

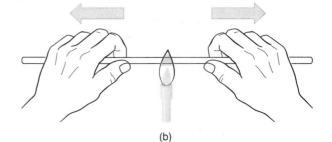

(b)

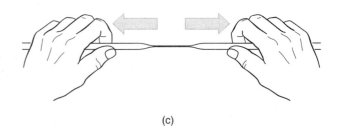

(c)

Figure 38   Bending glass tubing: (a) burner with wing top, showing suitable flame for bending glass tubing; (b) heating glass tubing to soften it; (c) bending glass tubing

Figure 39   Preparing glass droppers or pipets: (a) heating glass tubing; (b) pulling tubing in the flame; (c) pulling tubing out of the flame

Appendix B: Reference Tables

A Periodic Table is on the inside front cover. On the inside back cover is a table of element names, symbols, atomic numbers, and atomic masses.

Table 1 Physical Constants

Quantity	Symbol	Traditional Units	SI units
atomic mass unit (1/12 the mass of ^{12}C atom)	amu *or* u	1.6605×10^{-24} g	1.6605×10^{-27} kg
Avogadro's number	N	6.0221367×10^{23} particles/mol	6.0221367×10^{23} particles/mol
Bohr radius	a_0	0.52918 Å 5.2918×10^{-9} cm	5.2918×10^{-11} m
Boltzmann constant	k	1.3807×10^{-16} erg/K	1.3807×10^{-23} J/K
charge-to-mass ratio of electron	e/m	1.7588×10^8 coulomb/g	1.7588×10^{11} coulomb/kg
electronic charge	e	1.6022×10^{-19} coulomb 4.8033×10^{-10} esu	1.6022×10^{-19} coulomb
Faraday constant	F	96,485 coulomb/mol e^- 23.06 kcal/volt · mol e^-	96,485 coulomb/mol e^- 96,485 joules/volt · mol e^-
gas constant	R	0.08206 L atm/mol · K 1.987 cal/mol · K	8.3145 Pa dm^3/mol · K 8.3145 joules/mol · K
molar volume (STP)	V_m	22.414 L/mol	22.414×10^{-3} m^3/mol 22.414 dm^3/mol
Planck's constant	h	6.6261×10^{-27} erg · s	$6.6260755 \times 10^{-34}$ joule · s
Rydberg constant	R_α	3.289×10^{15} cycles/s 2.1799×10^{-11} erg	1.0974×10^7 m^{-1} 2.1799×10^{-18} joule
velocity of light (in a vacuum)	c	2.9979×10^{10} cm/s 186,282 miles/s	2.9979×10^8 m/s

Table 2 Conversion Tables

Length	
SI unit: meter (m)	
1 kilometer (km)	= 1000. meters
	= 0.6215 mile
1 meter (m)	= 100. centimeters
1 centimeter (cm)	= 10. millimeters(mm)
1 nanometer (nm)	= 1.00×10^{-9} meter
1 picometer (pm)	= 1.00×10^{-12} meter
1 inch (in.)	= 2.54 centimeter (exactly)
1 Ångstrom (Å)	= 1.00×10^{-10} meter

Volume	
SI unit: cubic meter (m³)	
1 liter (L)	= 1.00×10^{-3} m³
	= 1000. cm³
	= 1.056710 quarts
1 gallon	= 4.00 quarts

Pressure	
SI unit: pascal (Pa)	
1 pascal (Pa)	= 1 N/m²
	= 1 kg/m · s²
1 atmosphere (atm)	= 101.325 kilopascals
	= 760. mm Hg
	= 760 torr
	= 14.70 lb/in.²

Mass	
SI unit: kilogram (kg)	
1 kilogram	= 1000. grams
1 gram	= 1000. milligrams
1 pound	= 453.59237 grams
	= 16 ounces
1 ton	= 2000. pounds

Temperature	
SI unit: kelvin (K)	
0 K	= –273.15 °C
K	= °C + 273.15 °C
°C conversion	= 5/9(°F – 32)
°F conversion	= 9/5(°C) + 32

Energy	
SI unit: joule (J)	
1 joule	= 1 kg · m²/s²
	= 0.23901 calorie
	= 1 C × 1 V
1 calorie	= 4.184 joules

Table 3   Common Acids and Bases

	Strong Electrolytes		Weak Electrolytes	
Strong acids			*Weak acids*	
	HCl	hydrochloric acid	H_3PO_4	phosphoric acid
	HNO_3	nitric acid	CH_3CO_2H	acetic acid
	$HClO_4$	perchloric acid	H_2CO_3	carbonic acid
	H_2SO_4	sulfuric acid		
Strong bases			*Weak base*	
	NaOH	sodium hydroxide	NH_3	ammonia
	KOH	potassium hydroxide		
	$Ca(OH)_2$	calcium hydroxide		

Table 4   General Solubility Rules for Ionic Compounds in Water

Soluble Compounds	Exceptions to the Rule
Most compounds containing Na^+, K^+, or NH_4^+ ions	
Most compounds containing NO_3^-, $C_2H_3O_2^-$, ClO_3^-, or ClO_4^- ions	
Most compounds containing Cl^-, Br^-, or I^- ions	Halides containing Ag^+, Pb^{2+}, or Hg_2^{2+} ions are insoluble
Most compounds containing F^- ions	Fluorides containing Mg^{2+}, Ca^{2+}, Sr^{2+}, Ba^{2+}, or Pb^{2+} ions are insoluble
Most compounds containing SO_4^{2-} ions	$BaSO_4$ and $CaSO_4$ are insoluble

Insoluble Compounds	Exceptions to the Rule
Most compounds containing Ag^+, Pb^{2+}, or Hg_2^{2+} ions	
Most compounds containing carbonate, CO_3^{2-}, phosphate, PO_4^{3-}, oxalate, $C_2O_4^{2-}$, chromate, CrO_4^{2-}, sulfide, S^{2-}, hydroxide, OH^-, or oxide, O^{2-} ions	Most compounds containing Na^+, K^+, or NH_4^+ ions are soluble

Table 5   Vapor Pressure and Temperature Data

Temperature, °C	Vapor Pressure, torr	Temperature, °C	Vapor Pressure, torr
0	4.6	23	21.1
5	6.5	24	22.4
10	9.2	25	23.8
11	9.8	26	25.2
12	10.5	27	26.7
13	11.2	28	28.3
14	12.0	29	30.0
15	12.8	30	31.8
16	13.6	31	33.7
17	14.5	32	35.7
18	15.5	33	37.7
19	16.5	34	39.9
20	17.5	35	42.2
21	18.7	–	–
22	19.8	100	760.0

Table 6   Standard Reduction Potentials, $E°$, in Aqueous Solution, at 25 °C

Acidic Solution	Standard Reduction Potential, $E°$ (volts)
$F_2(g) + 2\,e^- \rightarrow 2\,F^-(aq)$	2.87
$Co^{3+}(aq) + e^- \rightarrow Co^{2+}(aq)$	1.82
$Pb^{4+}(aq) + 2\,e^- \rightarrow Pb^{2+}(aq)$	1.8
$H_2O_2(aq) + 2\,H^+(aq) + 2\,e^- \rightarrow 2\,H_2O$	1.77
$NiO_2(s) + 4H^+(aq) + 2\,e^- \rightarrow Ni^{2+}(aq) + 2\,H_2O$	1.7
$PbO_2(s) + SO_4^{2-}(aq) + 4\,H^+(aq) + 2\,e^- \rightarrow PbSO_4(s) + 2\,H_2O$	1.685
$Au^+(aq) + e^- \rightarrow Au(s)$	1.68
$2\,HClO(aq) + 2\,H^+(aq) + 2\,e^- \rightarrow Cl_2(g) + 2\,H_2O$	1.63
$Ce^{4+}(aq) + e^- \rightarrow Ce^{3+}(aq)$	1.61
$NaBiO_3(s) + 6\,H^+(aq) + 2\,e^- \rightarrow Bi^{3+}(aq) + Na^+(aq) + 3\,H_2O$	~1.6
$MnO_4^-(aq) + 8\,H^+(aq)\ 5\,e^- \rightarrow Mn^{2+}(aq) + 4\,H_2O$	1.51
$Au^{3+}(aq) + 3\,e^- \rightarrow Au(s)$	1.50
$ClO_3^-(aq) + 6\,H^+(aq) + 5\,e^- \rightarrow 1/2\,Cl_2(g) + 3\,H_2O$	1.47
$BrO_3^-(aq) + 6\,H^+(aq) + 6\,e^- \rightarrow Br^-(aq) + 3\,H_2O$	1.44
$Cl_2(g) + 2\,e^- \rightarrow 2\,Cl^-(aq)$	1.358
$Cr_2O_7^{2-}(aq) + 14\,H^+(aq) + 6\,e^- \rightarrow 2\,Cr^{3+}(aq) + 7\,H_2O$	1.33
$N_2H_5^+(aq) + 3\,H^+(aq) + 2\,e^- \rightarrow 2\,NH_4^+(aq)$	1.24
$MnO_2(s) + 4\,H^+(aq) + 2\,e^- \rightarrow Mn^{2+}(aq) + 2\,H_2O$	1.23
$O_2(g) + 4\,H^+(aq) + 4\,e^- \rightarrow 2\,H_2O$	1.229
$Pt^{2+}(aq) + 2\,e^- \rightarrow Pt(s)$	1.2
$IO_3^-(aq) + 6\,H^+(aq) + 5\,e^- \rightarrow 1/2\,I_2(aq) + 3\,H_2O$	1.195
$ClO_4^-(aq) + 2\,H^+(aq) + 2\,e^- \rightarrow ClO_3^-(aq) + H_2O$	1.19
$Br_2(g) + 2\,e^- \rightarrow 2\,Br^-(aq)$	1.066
$AuCl_4^-(aq) + 3\,e^- \rightarrow Au(s) + 4\,Cl^-(aq)$	1.00
$Pd^{2+}(aq) + 2\,e^- \rightarrow Pd(s)$	0.987
$NO_3^-(aq) + 4\,H^+(aq) + 3\,e^- \rightarrow NO(g) + 2\,H_2O$	0.96
$2\,Hg^{2+}(aq) + 2\,e^- \rightarrow Hg_2^{2+}(aq)$	0.920
$Hg^{2+}(aq) + 2\,e^- \rightarrow Hg(\ell)$	0.855
$Ag^+(aq) + e^- \rightarrow Ag(s)$	0.7994
$Hg_2^{2+}(aq) + 2\,e^- \rightarrow 2\,Hg(\ell)$	0.789
$Fe^{3+}(aq) + e^- \rightarrow Fe^{2+}(aq)$	0.771
$SbCl_6^-(aq) + 2\,e^- \rightarrow SbCl_4^-(aq) + 2\,Cl^-(aq)$	0.75
$[PtCl_4]^{2-}(aq) + 2\,e^- \rightarrow Pt(s) + 4\,Cl^-(aq)$	0.73
$O_2(g) + 2\,H^+(aq) + 2\,e^- \rightarrow H_2O_2(aq)$	0.682
$[PtCl_6]^{2-}(aq) + 2\,e^- \rightarrow [PtCl_4]^{2-}(aq) + 2\,Cl^-(aq)$	0.68
$H_3AsO_4(aq) + 2\,H^+(aq) + 2\,e^- \rightarrow H_3AsO_3(aq) + H_2O$	0.58
$I_2(s) + 2\,e^- \rightarrow 2\,I^-(aq)$	0.535
$TeO_2(s) + 4\,H^+(aq) + 4\,e^- \rightarrow Te(s) + 2\,H_2O$	0.529
$Cu^+(aq) + e^- \rightarrow Cu(s)$	0.521
$[RhCl_6]^{3-}(aq) + 3\,e^- \rightarrow Rh(s) + 6\,Cl^-(aq)$	0.44
$Cu^{2+}(aq) + 2\,e^- \rightarrow Cu(s)$	0.337
$HgCl_2(s) + 2\,e^- \rightarrow 2\,Hg(\ell) + 2\,Cl^-(aq)$	0.27
$AgCl(s) + e^- \rightarrow Ag(s) + Cl^-(aq)$	0.222
$SO_4^{2-}(aq) + 4\,H^+(aq) + 2\,e^- \rightarrow SO_2(g) + 2\,H_2O$	0.20
$SO_4^{2-}(aq) + 4\,H^+(aq) + 2\,e^- \rightarrow H_2SO_3(aq) + H_2O$	0.17
$Cu^{2+}(aq) + e^- \rightarrow Cu^+(aq)$	0.153
$Sn^{4+}(aq) + 2\,e^- \rightarrow Sn^{2+}(aq)$	0.15
$S(s) + 2\,H^+(aq) + 2\,e^- \rightarrow H_2S(aq)$	0.14
$AgBr(s) + e^- \rightarrow Ag(s) + Br^-(aq)$	0.0713
$2\,H^+(aq) + 2\,e^- \rightarrow H_2(g)$ (reference electrode)	0.0000
$N_2O(g) + 6\,H^+(aq) + H_2O + 4\,e^- \rightarrow 2\,NH_3OH^+(aq)$	−0.05

Acidic Solution	Standard Reduction Potential, $E°$ (volts)
$Pb^{2+}(aq) + 2\ e^- \rightarrow Pb(s)$	−0.126
$Sn^{2+}(aq) + 2\ e^- \rightarrow Sn(s)$	−0.14
$AgI(s) + e^- \rightarrow Ag(s) + I^-(aq)$	−0.15
$[SnF_6]^{2-}(aq) + 4\ e^- \rightarrow Sn(s)\ + 6\ F^-(aq)$	−0.25
$Ni^{2+}(aq) + 2\ e^- \rightarrow Ni(s)$	−0.25
$Co^{2+}(aq) + 2\ e^- \rightarrow Co(s)$	−0.28
$Tl^+(aq) + e^- \rightarrow Tl(s)$	−0.34
$PbSO_4(s) + 2\ e^- \rightarrow Pb(s) + SO_4^{2-}(aq)$	−0.356
$Se(s)\ + 2\ H^+(aq) + 2\ e^- \rightarrow H_2Se(aq)$	−0.40
$Cd^{2+}(aq) + 2\ e^- \rightarrow Cd(s)$	−0.403
$Cr^{3+}(aq) + e^- \rightarrow Cr^{2+}(aq)$	−0.41
$Fe^{2+}(aq) + 2\ e^- \rightarrow Fe(s)$	−0.44
$2\ CO_2(g) + 2\ H^+(aq) + 2\ e^- \rightarrow (COOH)_2(aq)$	−0.49
$Ga^{3+}(aq)\ + 3\ e^- \rightarrow Ga(s)$	−0.53
$HgS(s) + 2\ H^+(aq)\ + 2\ e^- \rightarrow Hg(\ell) + H_2S(g)$	−0.72
$Cr^{3+}(aq) + 3\ e^- \rightarrow Cr(s)$	−0.74
$Zn^{2+}(aq) + 2\ e^- \rightarrow Zn(s)$	−0.763
$Cr^{2+}(aq) + 2\ e^- \rightarrow Cr(s)$	−0.91
$FeS(s) + 2\ e^- \rightarrow Fe(s) + S^{2-}(aq)$	−1.01
$Mn^{2+}(aq) +\ 2\ e^- \rightarrow Mn(s)$	−1.18
$V^{2+}(aq) + 2\ e^- \rightarrow V(s)$	−1.18
$CdS(s) + 2\ e^- \rightarrow Cd(s) + S^{2-}(aq)$	−1.21
$ZnS(s) + 2\ e^- \rightarrow Zn(s) + S^{2-}(aq)$	−1.44
$Zr^{4+}(aq) + 4\ e^- \rightarrow Zr(s)$	−1.53
$Al^{3+}(aq)\ + 3\ e^- \rightarrow Al(s)$	−1.66
$H_2(g) + 2\ e^- \rightarrow 2\ H^-(aq)$	−2.25
$Mg^{2+}(aq) + 2\ e^- \rightarrow Mg(s)$	−2.37
$Na^+(aq) + e^- \rightarrow Na(s)$	−2.714
$Ca^{2+}(aq) + 2\ e^- \rightarrow Ca(s)$	−2.87
$Sr^{2+}(aq) + 2\ e^- \rightarrow Sr(s)$	−2.89
$Ba^{2+}(aq) + 2\ e^- \rightarrow Ba(s)$	−2.90
$Rb^+(aq) + e^- \rightarrow Rb(s)$	−2.925
$K^+(aq) + e^- \rightarrow K(s)$	−2.925
$Li^+(aq) + e^- \rightarrow Li(s)$	−3.045

Basic Solution	Standard Reduction Potential, $E°$ (volts)
$ClO^-(aq) + H_2O + 2\,e^- \rightarrow Cl^-(aq) + 2\,OH^-(aq)$	0.89
$OOH^-(aq) + H_2O + 2\,e^- \rightarrow 3\,OH^-(aq)$	0.88
$2\,NH_2OH(aq) + 2\,e^- \rightarrow N_2H_4(aq) + 2\,OH^-(aq)$	0.74
$ClO_3^-(aq) + 3\,H_2O + 6\,e^- \rightarrow Cl^-(aq) + 6\,OH^-(aq)$	0.62
$MnO_4^-(aq) + 2\,H_2O + 3\,e^- \rightarrow MnO_2(s) + 4\,OH^-(aq)$	0.588
$MnO_4^-(aq) + e^- \rightarrow MnO_4^{2-}(aq)$	0.564
$NiO_2(s) + 2\,H_2O + 2\,e^- \rightarrow Ni(OH)_2(s) + 2\,OH^-(aq)$	0.49
$Ag_2CrO_4(s) + 2\,e^- \rightarrow 2\,Ag(s) + CrO_4^{2-}(aq)$	0.446
$O_2(g) + 2\,H_2O + 4\,e^- \rightarrow 4\,OH^-(aq)$	0.40
$ClO_4^-(aq) + H_2O + 2\,e^- \rightarrow ClO_3^-(aq) + 2\,OH^-(aq)$	0.36
$Ag_2O(s) + H_2O + 2\,e^- \rightarrow 2\,Ag(s) + 2\,OH^-(aq)$	0.34
$2\,NO_2^-(aq) + 3\,H_2O + 4\,e^- \rightarrow N_2O(g) + 6\,OH^-(aq)$	0.15
$N_2H_4(aq) + 2\,H_2O + 2\,e^- \rightarrow 2\,NH_3(aq) + 2\,OH^-(aq)$	0.10
$[Co(NH_3)_6]^{3+}(aq) + e^- \rightarrow [Co(NH_3)_6]^{2+}(aq)$	0.10
$HgO(s) + H_2O + 2\,e^- \rightarrow Hg(\ell) + 2\,OH^-(aq)$	0.0984
$O_2(g) + H_2O + 2\,e^- \rightarrow OOH^-(aq) + OH^-(aq)$	0.076
$NO_3^-(aq) + H_2O + 2\,e^- \rightarrow NO_2^-(aq) + 2\,OH^-(aq)$	0.01
$MnO_2(s) + 2\,H_2O + 2\,e^- \rightarrow Mn(OH)_2(s) + 2\,OH^-(aq)$	−0.05
$CrO_4^{2-}(aq) + 4\,H_2O + 3\,e^- \rightarrow Cr(OH)_3(s) + 5\,OH^-(aq)$	−0.12
$Cu(OH)_2(s) + 2\,e^- \rightarrow Cu(s) + 2\,OH^-(aq)$	−0.36
$S(s) + 2\,e^- \rightarrow S^{2-}(aq)$	−0.48
$Fe(OH)_3(s) + e^- \rightarrow Fe(OH)_2(s) + OH^-(aq)$	−0.56
$2\,H_2O + 2\,e^- \rightarrow H_2(g) + 2\,OH^-(aq)$	−0.8277
$2\,NO_3^-(aq) + 2\,H_2O + 2\,e^- \rightarrow N_2O_4(g) + 4\,OH^-(aq)$	−0.85
$Fe(OH)_2(s) + 2\,e^- \rightarrow Fe(s) + 2\,OH^-(aq)$	−0.877
$SO_4^{2-}(aq) + H_2O + 2\,e^- \rightarrow SO_3^{2-}(aq) + 2\,OH^-(aq)$	−0.93
$N_2(g) + 4\,H_2O + 4\,e^- \rightarrow N_2H_4(aq) + 4\,OH^-(aq)$	−1.15
$[Zn(OH)_4]^{2-}(aq) + 2\,e^- \rightarrow Zn(s) + 4\,OH^-(aq)$	−1.22
$Zn(OH)_2(s) + 2\,e^- \rightarrow Zn(s) + 2\,OH^-(aq)$	−1.245
$[Zn(CN)_4]^{2-}(aq) + 2\,e^- \rightarrow Zn(s) + 4\,CN^-(aq)$	−1.26
$Cr(OH)_3(s) + 3\,e^- \rightarrow Cr(s) + 3\,OH^-(aq)$	−1.30
$SiO_3^{2-}(aq) + 3\,H_2O + 4\,e^- \rightarrow Si(s) + 6\,OH^-(aq)$	−1.70

Table 7 Acid Ionization Constants, K_a, at 25 °C

Acid	Formula and Ionization Equation	K_a
Acetic	$CH_3COOH \rightleftharpoons H^+ + CH_3COO^-$	1.8×10^{-5}
Arsenic	$H_3AsO_4 \rightleftharpoons H^+ + H_2AsO_4^-$	$K_1 = 2.5 \times 10^{-4}$
	$H_2AsO_4^- \rightleftharpoons H^+ + HAsO_4^{2-}$	$K_2 = 5.6 \times 10^{-8}$
	$HAsO_4^{2-} \rightleftharpoons H^+ + AsO_4^{3-}$	$K_3 = 3.0 \times 10^{-13}$
Arsenous	$H_3AsO_3 \rightleftharpoons H^+ + H_2AsO_3^-$	$K_1 = 6.0 \times 10^{-10}$
	$H_2AsO_3^- \rightleftharpoons H^+ + HAsO_3^{2-}$	$K_2 = 3.0 \times 10^{-14}$
Benzoic	$C_6H_5COOH \rightleftharpoons H^+ + C_6H_5COO^-$	6.3×10^{-5}
Boric	$H_3BO_3 \rightleftharpoons H^+ + H_2BO_3^-$	$K_1 = 7.3 \times 10^{-10}$
	$H_2BO_3^- \rightleftharpoons H^+ + HBO_3^{2-}$	$K_2 = 1.8 \times 10^{-13}$
	$HBO_3^{2-} \rightleftharpoons H^+ + BO_3^{3-}$	$K_3 = 1.6 \times 10^{-14}$
Carbonic	$H_2CO_3 \rightleftharpoons H^+ + HCO_3^-$	$K_1 = 4.2 \times 10^{-7}$
	$HCO_3^- \rightleftharpoons H^+ + CO_3^{2-}$	$K_2 = 4.8 \times 10^{-11}$
Citric	$H_3C_6H_5O_7 \rightleftharpoons H^+ + H_2C_6H_5O_7^-$	$K_1 = 7.4 \times 10^{-3}$
	$H_2C_6H_5O_7^- \rightleftharpoons H^+ + HC_6H_5O_7^{2-}$	$K_2 = 1.7 \times 10^{-5}$
	$HC_6H_5O_7^{2-} \rightleftharpoons H^+ + C_6H_5O_7^{3-}$	$K_3 = 4.0 \times 10^{-7}$
Cyanic	$HOCN \rightleftharpoons H^+ + OCN^-$	3.5×10^{-4}
Formic	$HCOOH \rightleftharpoons H^+ + HCOO^-$	1.8×10^{-4}
Hydrazoic	$HN_3 \rightleftharpoons H^+ + N_3^-$	1.9×10^{-5}
Hydrocyanic	$HCN \rightleftharpoons H^+ + CN^-$	4.0×10^{-10}
Hydrofluoric	$HF \rightleftharpoons H^+ + F^-$	7.2×10^{-4}
Hydrogen peroxide	$H_2O_2 \rightleftharpoons H^+ + HO_2^-$	2.4×10^{-12}
Hydrosulfuric	$H_2S \rightleftharpoons H^+ + HS^-$	$K_1 = 1.0 \times 10^{-7}$
	$HS^- \rightleftharpoons H^+ + S^{2-}$	$K_2 = 1.3 \times 10^{-13}$
Hypobromous	$HOBr \rightleftharpoons H^+ + OBr^-$	2.5×10^{-9}
Hypochlorous	$HOCl \rightleftharpoons H^+ + OCl^-$	3.5×10^{-8}
Nitrous	$HNO_2 \rightleftharpoons H^+ + NO_2^-$	4.5×10^{-4}
Oxalic	$H_2C_2O_4 \rightleftharpoons H^+ + HC_2O_4^-$	$K_1 = 5.9 \times 10^{-2}$
	$HC_2O_4^- \rightleftharpoons H^+ + C_2O_4^{2-}$	$K_2 = 6.4 \times 10^{-5}$
Phenol	$HC_6H_5O \rightleftharpoons H^+ + C_6H_5O^-$	1.3×10^{-10}
Phosphoric	$H_3PO_4 \rightleftharpoons H^+ + H_2PO_4^-$	$K_1 = 7.5 \times 10^{-3}$
	$H_2PO_4^- \rightleftharpoons H^+ + HPO_4^{2-}$	$K_2 = 6.2 \times 10^{-8}$
	$HPO_4^{2-} \rightleftharpoons H^+ + PO_4^{3-}$	$K_3 = 3.6 \times 10^{-13}$
Phosphorus	$H_3PO_3 \rightleftharpoons H^+ + H_2PO_3^-$	$K_1 = 1.6 \times 10^{-2}$
	$H_2PO_3^- \rightleftharpoons H^+ + HPO_3^{2-}$	$K_2 = 7.0 \times 10^{-7}$
Selenic	$H_2SeO_4 \rightleftharpoons H^+ + HSeO_4^-$	$K_1 = $ very large
	$HSeO_4^- \rightleftharpoons H^+ + SeO_4^{2-}$	$K_2 = 1.2 \times 10^{-2}$
Selenous	$H_2SeO_3 \rightleftharpoons H^+ + HSeO_3^-$	$K_1 = 2.7 \times 10^{-3}$
	$HSeO_3^- \rightleftharpoons H^+ + SeO_3^{2-}$	$K_2 = 2.5 \times 10^{-7}$
Sulfuric	$H_2SO_4 \rightleftharpoons H^+ + HSO_4^-$	$K_1 = $ very large
	$HSO_4^- \rightleftharpoons H^+ + SO_4^{2-}$	$K_2 = 1.2 \times 10^{-2}$
Sulfurous	$H_2SO_3 \rightleftharpoons H^+ + HSO_3^-$	$K_1 = 1.7 \times 10^{-2}$
	$HSO_3^- \rightleftharpoons H^+ + SO_3^{2-}$	$K_2 = 6.4 \times 10^{-8}$
Tellurous	$H_2TeO_3 \rightleftharpoons H^+ + HTeO_3^-$	$K_1 = 2 \times 10^{-3}$
	$HTeO_3^- \rightleftharpoons H^+ + TeO_3^{2-}$	$K_2 = 1 \times 10^{-8}$

Table 8   Base Ionization Constants, K_b, at 25 °C

Base	Formula and Ionization Equation	K_b
Ammonia	$NH_3 + H_2O \rightleftharpoons NH_4^+ + OH^-$	1.8×10^{-5}
Aniline	$C_6H_5NH_2 + H_2O \rightleftharpoons C_6H_5NH_3^+ + OH^-$	4.0×10^{-10}
Dimethylamine	$(CH_3)_2NH + H_2O \rightleftharpoons (CH_3)_2NH_2^+ + OH^-$	7.4×10^{-4}
Ethylenediamine	$(CH_2)_2(NH_2)_2 + H_2O \rightleftharpoons (CH_2)_2(NH_2)_2H^+ + OH^-$	$K_1 = 8.5 \times 10^{-5}$
	$(CH_2)_2(NH_2)_2H^+ + H_2O \rightleftharpoons (CH_2)_2(NH_2)_2H_2^{2+} + OH^-$	$K_2 = 2.7 \times 10^{-8}$
Hydrazine	$N_2H_4 + H_2O \rightleftharpoons N_2H_5^+ + OH^-$	$K_1 = 8.5 \times 10^{-7}$
	$N_2H_5^+ + H_2O \rightleftharpoons N_2H_6^{2+} + OH^-$	$K_2 = 8.9 \times 10^{-16}$
Hydroxylamine	$NH_2OH + H_2O \rightleftharpoons NH_3OH^+ + OH^-$	6.6×10^{-9}
Methylamine	$CH_3NH_2 + H_2O \rightleftharpoons CH_3NH_3^+ + OH^-$	5.0×10^{-4}
Pyridine	$C_5H_5N + H_2O \rightleftharpoons C_5H_5NH^+ + OH^-$	1.5×10^{-9}
Trimethylamine	$(CH_3)_3N + H_2O \rightleftharpoons (CH_3)_3NH^+ + OH^-$	7.4×10^{-5}

Table 9 Solubility Product Constants, K_{sp}, at 25 °C

Substance	K_{sp}	Substance	K_{sp}
Aluminum compounds		Copper compounds	
$AlAsO_4$	1.6×10^{-16}	CuBr	5.3×10^{-9}
$Al(OH)_3$	1.9×10^{-33}	CuCl	1.9×10^{-7}
$AlPO_4$	1.3×10^{-20}	CuCN	3.2×10^{-20}
Antimony compounds		Cu_2O ($Cu^+ + OH^-$)	1.0×10^{-14}
Sb_2S_3	1.6×10^{-93}	CuI	5.1×10^{-12}
Barium compounds		Cu_2S	1.6×10^{-48}
$Ba_3(AsO_4)_2$	1.1×10^{-13}	CuSCN	1.6×10^{-11}
$BaCO_3$	8.1×10^{-9}	$Cu_3(AsO_4)_2$	7.6×10^{-36}
$BaC_2O_4 \cdot 2 H_2O$	1.1×10^{-7}	$CuCO_3$	2.5×10^{-10}
$BaCrO_4$	2.0×10^{-10}	$Cu_2[Fe(CN)_6]$	1.3×10^{-16}
BaF_2	1.7×10^{-6}	$Cu(OH)_2$	1.6×10^{-19}
$Ba(OH)_2 \cdot 8 H_2O$	5.0×10^{-3}	CuS	8.7×10^{-36}
$Ba_3(PO_4)_2$	1.3×10^{-29}	Gold compounds	
$BaSeO_4$	2.8×10^{-11}	AuBr	5.0×10^{-17}
$BaSO_3$	8.0×10^{-7}	AuCl	2.0×10^{-13}
$BaSO_4$	1.1×10^{-10}	AuI	1.6×10^{-23}
Bismuth compounds		$AuBr_3$	4.0×10^{-36}
BiOCl	7.0×10^{-9}	$AuCl_3$	3.2×10^{-25}
BiO(OH)	1.0×10^{-12}	$Au(OH)_3$	1×10^{-53}
$Bi(OH)_3$	3.2×10^{-40}	AuI_3	1.0×10^{-46}
BiI_3	8.1×10^{-19}	Iron compounds	
$BiPO_4$	1.3×10^{-23}	$FeCO_3$	3.5×10^{-11}
Bi_2S_3	1.6×10^{-72}	$Fe(OH)_2$	7.9×10^{-15}
Cadmium compounds		FeS	4.9×10^{-18}
$Cd_3(AsO_4)_2$	2.2×10^{-32}	$Fe_4[Fe(CN)_6]_3$	3.0×10^{-41}
$CdCO_3$	2.5×10^{-14}	$Fe(OH)_3$	6.3×10^{-38}
$Cd(CN)_2$	1.0×10^{-8}	Fe_2S_3	1.4×10^{-88}
$Cd_2[Fe(CN)_6]$	3.2×10^{-17}	Lead compounds	
$Cd(OH)_2$	1.2×10^{-14}	$Pb_3(AsO_4)_2$	4.1×10^{-36}
CdS	3.6×10^{-29}	$PbBr_2$	6.3×10^{-6}
Calcium compounds		$PbCO_3$	1.5×10^{-13}
$Ca_3(AsO_4)_2$	6.8×10^{-19}	$PbCl_2$	1.7×10^{-5}
$CaCO_3$	3.8×10^{-9}	$PbCrO_4$	1.8×10^{-14}
$CaCrO_4$	7.1×10^{-4}	PbF_2	3.7×10^{-8}
$CaC_2O_4 \cdot H_2O$	2.3×10^{-9}	$Pb(OH)_2$	2.8×10^{-16}
CaF_2	3.9×10^{-11}	PbI_2	8.7×10^{-9}
$Ca(OH)_2$	7.9×10^{-6}	$Pb_3(PO_4)_2$	3.0×10^{-44}
$CaHPO_4$	2.7×10^{-7}	$PbSeO_4$	1.5×10^{-7}
$Ca(H_2PO_4)_2$	1.0×10^{-3}	$PbSO_4$	1.8×10^{-8}
$Ca_3(PO_4)_2$	1.0×10^{-25}	PbS	8.4×10^{-28}
$CaSO_3 \cdot 2 H_2O$	1.3×10^{-8}	Magnesium compounds	
$CaSO_4 \cdot 2 H_2O$	2.4×10^{-5}	$Mg_3(AsO_4)_2$	2.1×10^{-20}
Chromium compounds		$MgCO_3 \cdot 3 H_2O$	4.0×10^{-5}
$CrAsO_4$	7.8×10^{-21}	MgC_2O_4	8.6×10^{-5}
$Cr(OH)_3$	6.7×10^{-31}	MgF_2	6.4×10^{-9}
$CrPO_4$	2.4×10^{-23}	$Mg(OH)_2$	1.5×10^{-11}
Cobalt compounds		$MgNH_4PO_4$	2.5×10^{-12}
$Co_3(AsO_4)_2$	7.6×10^{-29}	Manganese compounds	
$CoCO_3$	8.0×10^{-13}	$Mn_3(AsO_4)_2$	1.9×10^{-11}
$Co(OH)_2$	2.5×10^{-16}	$MnCO_3$	1.8×10^{-11}
$CoS(\alpha)$	5.9×10^{-21}	$Mn(OH)_2$	4.6×10^{-14}
$Co(OH)_3$	4.0×10^{-45}	MnS	5.1×10^{-15}
		$Mn(OH)_3$	$\sim 1 \times 10^{-36}$

Substance	K_{sp}	Substance	K_{sp}
Mercury compounds		AgI	1.5×10^{-16}
Hg_2Br_2	1.3×10^{-22}	Ag_3PO_4	1.3×10^{-20}
Hg_2CO_3	8.9×10^{-17}	Ag_2SO_3	1.5×10^{-14}
Hg_2Cl_2	1.1×10^{-18}	Ag_2SO_4	1.7×10^{-5}
Hg_2CrO_4	5.0×10^{-9}	Ag_2S	1.0×10^{-49}
Hg_2I_2	4.5×10^{-29}	AgSCN	1.0×10^{-12}
$Hg_2O \cdot H_2O$ $(Hg_2^{2+} + 2\ OH^-)$	1.6×10^{-23}	**Strontium compounds**	
Hg_2SO_4	6.8×10^{-7}	$Sr_3(AsO_4)_2$	1.3×10^{-18}
Hg_2S	5.8×10^{-44}	$SrCO_3$	9.4×10^{-10}
$Hg(CN)_2$	3.0×10^{-23}	$SrC_2O_4 \cdot 2\ H_2O$	5.6×10^{-8}
$Hg(OH)_2$	2.5×10^{-26}	$SrCrO_4$	3.6×10^{-5}
HgI_2	4.0×10^{-29}	$Sr(OH)_2 \cdot 8\ H_2O$	3.2×10^{-4}
HgS	3.0×10^{-53}	$Sr_3(PO_4)_2$	1.0×10^{-31}
Nickel compounds		$SrSO_3$	4.0×10^{-8}
$Ni_3(AsO_4)_2$	1.9×10^{-26}	$SrSO_4$	2.8×10^{-7}
$NiCO_3$	6.6×10^{-9}	**Tin compounds**	
$Ni(CN)_2$	3.0×10^{-23}	$Sn(OH)_2$	2.0×10^{-26}
$Ni(OH)_2$	2.8×10^{-16}	SnI_2	1.0×10^{-4}
NiS (α)	3.0×10^{-21}	SnS	1.0×10^{-28}
NiS (β)	1.0×10^{-26}	$Sn(OH)_4$	1×10^{-57}
NiS (γ)	2.0×10^{-28}	SnS_2	1×10^{-70}
Silver compounds		**Zinc compounds**	
Ag_3AsO_4	1.1×10^{-20}	$Zn_3(AsO_4)_2$	1.1×10^{-27}
AgBr	3.3×10^{-13}	$ZnCO_3$	1.5×10^{-11}
Ag_2CO_3	8.1×10^{-12}	$Zn(CN)_2$	8.0×10^{-12}
AgCl	1.8×10^{-10}	$Zn_3[Fe(CN)_6]$	4.1×10^{-16}
Ag_2CrO_4	9.0×10^{-12}	$Zn(OH)_2$	4.5×10^{-17}
AgCN	1.2×10^{-16}	$Zn_3(PO_4)_2$	9.1×10^{-33}
$Ag_4[Fe(CN)_6]$	1.6×10^{-41}	ZnS	1.1×10^{-21}
Ag_2O $(Ag^+ + OH^-)$	2.0×10^{-8}		

© 2002 Wadsworth Group

Appendix C:
General Laboratory References

Mahn, W.J. *Academic Laboratory Chemical Hazards Guidebook;* Van Nostrand Reinhold: New York, 1991.

Mahn, W.J. *Fundamentals of Laboratory Safety: Physical Hazards in the Academic Laboratory;* Van Nostrand Reinhold: New York, 1991.

CRC Handbook of Chemistry and Physics, 81st ed.; Lide, D.R., *Ed.;* CRC Press: Boca Raton, FL, 2000.

Handbook of Laboratory Safety, 5th ed.; Furr, A.K., *Ed.;* CRC Press: Boca Raton, FL, 2000.

Material Safety Data Sheets (MSDS)

Merck Index of Chemicals and Drugs, 12th ed.; Merck and Co.: Rahway, NJ, 1990.

Prudent Practices in the Laboratory: Handling and Disposal of Chemicals; National Academic Press: Washington, DC, 1995.

Safety in Academic Chemistry Laboratories, 5th ed.; American Chemical Society: Washington, DC, 1990.

The Sigma–Aldrich Library of Chemical Safety Data, 3rd ed.; Lenga, R.E., *Ed.;* Sigma–Aldrich Corp.: Milwaukee, WI, 1998.

Index

abscissa, 84
absorption of light, 59
accuracy, 81
acids, concentrated,
   adding to water, 33
acids, common, table of, 95
analytical balance, 17
analytical wavelength, 59
aspirator, water, 47

balances, 15
   analytical, 17
   four beam, 16
   general rules for, 15
   leveling, 15
   top-loading, 15
   triple-beam, 16
bases, common, table of, 95
beaker, 10
Beral pipet, 11, 32, 48, 53
best straight line, 84
blank, 61
blanket, fire, 6
Büchner funnel, 10, 47
Bunsen burner, 11, 37
buret, 10, 24
   cleaning, 89
   control devices, 24
   filling, 24
   reading, 25
burns, 6

calibration marks
   on burets, 25
   on pipets, 22
   on volumetric flasks, 26
centrifuge, 48
chamber, weighing, 18
clay triangle, 11, 40
cleaning glassware, 89
clothing, suitable, 2
constants, physical, table of, 93
contact lenses, 1
control bar, 15, 18
conversion,
   % T to A, 61
   tables, 94
correlation, 85
crucible, 10
crucible tongs, 11
cuvet, 48
cylinder, graduated, 10, 22

data
   crossing out, 69
   recording, 69
data table, 83

decantation, 45
dependent variable, 83
desiccator, 11
detecting odors, 2
digit term, 73
discarding containers, 4
disposal, chemical, 3, 4
draft shield, 18
dropper bottles, 32
droppers, preparing, 91

$E°$, 96–98
electrode
   combination, 65
   indicator, 65
   reference, 65
elements, table of, *inside back cover*
equipment, illustrations of, 11
Erlenmeyer flask, 10
error
   random, 81
   systematic, 81
evaporating dish, 10
exponential notation, 73
exponential term, 73
extrapolation, 83

figures in notebook, 69
figures, significant, 77
file, triangular, 11, 89
filter flask, 10, 47
filter paper, folding of, 46
filtrate, 46
filtration, 45
   gravity, 45
   suction, 47
firepolishing glassware, 89
fire extinguishers, 6
fire safety, 6
flame
   oxidizing (nonluminous), 38
   reducing (luminous), 38
flame spreader, 11, 89
flask
   Erlenmeyer, 10
   filter, 10, 47
   Florence, 10
   volumetric, 10, 26
four beam balance, 16
fume hood, 2
funnel
   Büchner, 10, 47
   conical, 10, 45
   short-stem, *see* conical

gas cock, 38
glass, broken, 3

glass rod, 10, 46
glass tubing
   bending, 89
   cutting, 89
   firepolishing, 89
   inserting into stoppers, 4
glassware
   chipped or cracked, 3
   cleaning, 89
   illustrations of, 10
goggles, 1
graduated cylinder, 10, 22
graphs, 83
gravity filtration, 45

hazards scale, 3
heating block, aluminum, 11, 53
heating
   liquids, 39
   solids, 40
   to constant mass, 40
hot objects and burns, 3
hot plates, 11, 38, 55
hot-water bath, 39

independent variable, 83
ionization constants, tables of
   acids, 99
   bases, 100

K_a, 99
K_b, 100
K_{sp}, 101–102

laboratory notebook, 69
labeling
   beakers, 31
   centrifuge tubes, 48
   reading, 3
light, visible, 59
line graph, 83
linear regression line, 85
liquids
   diluting, 33
   flammable, 4, 39
   mixing, 33
   transferring, 32
lubrication
   of glassware, 4
   of stopcock, 24
luminous flame, 83

mass determination, 15
masses, sliding, 16
Material Safety Data Sheets
   (MSDS), 4
mean, calculating, 81